AF477466

NESTING BIOLOGY
AND ASSOCIATES OF *MELITOMA*
(Hymenoptera, Anthophoridae)

Nesting Biology and Associates of *Melitoma*

(Hymenoptera, Anthophoridae)

by E. G. Linsley, J. W. MacSwain, and C. D. Michener

UNIVERSITY OF CALIFORNIA PRESS
Berkeley • Los Angeles • London

UNIVERSITY OF CALIFORNIA PUBLICATIONS IN ENTOMOLOGY

Volume 90

Issue Date: August 1980

UNIVERSITY OF CALIFORNIA PRESS
BERKELEY AND LOS ANGELES, CALIFORNIA

UNIVERSITY OF CALIFORNIA PRESS, LTD.
LONDON, ENGLAND

Library of Congress Cataloging in Publication Data

Linsley, Earle Gorton, 1910-
Nesting biology and associates of Melitoma (Hymenoptera, Anthophoridae)

University of California publications in entomology; v. 90)
Includes bibliographical references.
1. Melitoma—Behavior. 2. Melitoma—Ecology. 3. Nest building. 4. Symbiosis. 5. Insects—Behavior. 6. Insects—Ecology. I. MacSwain, John Winslow, 1915-1970 joint author. II. Michener, Charles Duncan, 1918- joint author. III. Title. IV. Series: California. University. University of California publications in entomology; v. 90.
QL568.A53L56 595.79'9 80-12433

ISBN 0-520-09618-5
LIBRARY OF CONGRESS CATALOG CARD NUMBER: 80-12433

PRINTED IN THE UNITED STATES OF AMERICA

Contents

Acknowledgments

The studies in Mexico by E.G.L. and J.W.MacS. were parts of a series of investigations on the habits of anthophorid bees in Mexico made possible by grants-in-aid from the Associates in Tropical Biogeography, University of California. The studies in Colombia by C.D.M. were possible thanks to M. J. West-Eberhard and W. G. Eberhard of the Universidad del Valle, Cali. Those in French Guiana were made possible by facilities maintained by Gard Otis, David and Gretchen Roubik, Penelope F. Kukuk, and Mark Winston for work on the Africanized honeybee, under the direction of O. R. Taylor (U.S. Dept. of Agriculture contract 12-14-7001-363), and by funding through the University of Kansas Endowment Association.

E.G.L. and J.W.MacS. acknowledge especially the collaboration of Ray F. Smith in most of the Mexican investigations, and the field assistance of C. D. MacNeill, R. C. Bechtel, E. I. Schlinger, and D. D. Linsdale in the Chiapas phases of this study. We are indebted to R. B. Selander, who kindly made available extensive data from material collected with the aid of Mrs. Selander at Tuxtla Gutiérrez, Chiapas, in the summer of 1955, while working under grants from the American Philosophical Society and the Society of Sigma Xi. Dr. Selander also read an early version of the manuscript and offered valuable suggestions, as did G. E. Bohart and P. D. Hurd, Jr.

Preparation of the manuscript at the University of Kansas was facilitated by National Science Foundation Grants DEB77-23035 and BMS78-07707, and by Joetta Weaver, who carefully edited and typed it.

For identification of various associates of *Melitoma* we are indebted to a number of specialists, as follows: *Cymatodera,* W. F. Barr; *Dasymutilla,* C. E. Mickel; *Monodontomerus,* B. D. Burks; *Anthrax,* F. C. Cole; *Chaetodactylus,* H. H. J. Nesbitt; *Chalicodoma* and *Coelioxys,* T. B. Mitchell; and fungi, the late E. A. Steinhaus and P. L. Lentz. Dr. T. J. Zavortink of the University of San Francisco kindly examined some of the specimens of *Melitoma* to verify identifications, and provided dates of capture of museum specimens from many tropical sites to supplement information on seasonality derived from nesting sites. We are also indebted to him for information on the status of the four species discussed.

INTRODUCTION

Melitoma segmentaria (formerly called *M. euglossoides* and *M. fulvifrons*; see Moure, 1960) ranges from Argentina to the Mexican Plateau. In Central America and Mexico it is largely replaced by the very similar *M. marginella*, which also occurs in southern Texas. These forms are closely related to two north temperate species: *M. grisella*, found in dry plains from Texas and New Mexico to South Dakota, and *M. taurea*, found in the more mesic eastern United States from Florida to New Jersey, west to Texas and Kansas. These four species, which have sometimes been placed in the synonymous genus *Entechnia*, and which are similar enough that their specific distinctness has been doubted, together range across some 80 degrees of latitude. They occur from sea level in both temperate and tropical regions (Argentina, Uruguay, French Guiana, Trinidad) to altitudes of 2200 m (Mexican Plateau) and 3815 m (Puno, Peru). Habitats include the following and all intermediates: wet lowland tropics (French Guiana), the grasslands of Uruguay and Argentina, irrigated places in the desert (near Lima, Peru), the desert scrub of the Mexican Plateau, the short grass prairie of the high plains of the United States, and the mesic, formerly forested, eastern United States.

Information on the nesting of all four species listed above is summarized in this paper. In South America there are a few other related species, but except for one which is unidentified (Michener and Lange, 1958), no biological observations have been made concerning such species. They will not be treated in this paper.

We are indebted to Dr. Thomas J. Zavortink, who plans to revise the genus *Melitoma*, for information on these species. He reports that the males have specific genitalic characters. The females are morphologically similar, however, and those of *M. segmentaria* and *M. marginella* differ, so far as is known, only in the color of the scape (brown or black in *marginella*, orange or red in *segmentaria*)—an attribute which is not always reliable, since there are some populations of *segmentaria* with the scape dark.

All the *Melitoma* forms discussed in the present paper collect pollen primarily if not exclusively from flowers of *Ipomoea* (Convolvulaceae). Bodkin (1918) and Michener and Lange (1958) reported *M. segmentaria* at flowers of cultivated cotton in Guyana and western Peru, respectively, but did not verify that the bees were collecting pollen.

Some of the habits of *M. segmentaria* have been briefly discussed by Doello-Jurado (1912), Bertoni (1918, 1929), Linsley, MacSwain, and Smith (1956), Michener, Lange, Bigarella, and Salamuni (1958), Michener and Lange (1958), and Michener (1975). Torchio (1974) discusses a site inhabited by *M. marginella*. Material on the biology of

M. grisella has been recorded by Hungerford and Williams (1912), Cockerell (1934a and b), and Linsley (1960). From the biological standpoint, the best-known species has been *M. taurea,* thanks to observations by Say (1837, quoted also by Ashmead, 1894, and treated in a separate account by Ashmead in the same paper under the name *Epeolus donatus,* Robertson (1914), Rau (1926, 1929, 1934), Chandler (1958), and Michener (1975). We have not attempted to list here the various records in the more-or-less taxonomic literature of capture of *Melitoma* species on flowers of *Ipomoea.* It suffices to note that the relationship of these bees to *Ipomoea* has long been known, and led to the specific name *ipomoeae,* a synonym of *segmentaria* (Schrottky, 1902). In most of the works listed above emphasis is on nesting habits or flower relationships; little has been reported on the predators, parasites, or inquilines associated with *Melitoma.* The purpose of the present paper is to review and supplement our knowledge of nesting biology and to direct attention to the associates in and around the nests of *Melitoma.* Our studies reported below were made in the Mexican states of Chiapas, Guanajuato, and Zacatecas; in the Provincia del Valle, Colombia; and in French Guiana.

The studies in Mexico and analyses of cells taken there were made by E.G.L. and J.W.MacS., who are almost exclusively responsible for the content of the sections on Competitors on Flowers, Organisms Associated with *Melitoma* Cells in Mexico, and Associates of Aggregating Anthophoridae in North America. C.D.M. made studies in Colombia and French Guiana, as well as previously described investigations in Brazil, Peru, and Missouri. A draft of the results of the Mexican studies was prepared by E.G.L. and J.W.MacS., and was supplemented and revised by C.D.M., who is largely responsible for the sections on the Seasonal Cycle and Nests.

NEST SITES

In the following comments on previously described nesting sites of *M. segmentaria*, together with descriptions of localities where we gathered data on that species and *M. marginella*, those probably or certainly attributable to *marginella* are indicated by a notation in square brackets at the end of the account. This material is followed by comments on sites recorded for *M. grisella* and *M. taurea*. All are numbered for convenient reference in the rest of the text. Sites 1 to 6 are in the south temperate zone, 7 and 8 in the southern part of the tropics, 9 and 10 in the equatorial region proper, 11 to 19 in the northern part of the tropics, and 20 to 30 in the north temperate zone. The sites described for the first time in this paper, and from which new data have been gathered, are 9, 10, 12–15, 17, and 18.

Melitoma segmentaria and *M. marginella*

1. *Buenos Aires, Argentina.* A single nest in a bank 1.5 m high in Parque Lezama (Doello-Jurado, 1912).

2. *Gualeguaychú, Entre Rios, Argentina.* About 25 nests in mortar between bricks of a wall. All were in an area of about 4 sq ft (about 0.6 m^2); the rest of the extensive wall was unoccupied (Doello-Jurado, 1912).

3. *Montevideo, Uruguay.* Nests in a north-facing bank (Brèthes, 1909).

4. *Puerto Bertoni, Paraguay.* Numerous nests in a clay oven, the interior wall of which had been converted over the years to a mass of cells (Bertoni, 1918). This site was in a lowland and primitively forested area on the Rio Paraná, altitude about 200 m.

5. *Puerto Alegre, Paraguay.* Nests in mud walls of a house, mixed with more numerous nests of *Centris lanipes* (Bertoni, 1929).

6. *Curitiba, Paraná, Brazil.* A total of 16 nests, mostly scattered but some in groups of 2 or 3, in two small, hard clay areas of extensive vertical roadside banks, mostly of softer material, in the suburb of Bariguí at an altitude of about 1000 m (Michener, Lange, Bigarella, and Salamuni, 1958; Michener and Lange, 1958). The banks inhabited by *Melitoma* faced north or northeast. One of the sites was under a protecting overhanging root.

7. *São Carlos, São Paulo, Brazil.* A few nests in a shaded wall of unbaked bricks (Michener and Lange, 1968).

8. *Near Lima, Peru.* Nests in adobe walls near a cultivated cotton field, the flowers of which were being visited by the bees in June 1955 (Michener and Lange, 1958).

9. *Cali, Provincia del Valle, Colombia.* Many nests, most of them unoccupied, were

found in the vertical and overhanging clay walls of an excavation (small mine entrance) under a huge boulder behind the Faculdad de Medicina in Barrio San Fernando, Cali, Colombia, altitude 1000 m. These walls were in deep shadow. This is the same site described by Michener (1974a) and Eberhard (1974). A few other nests were found in a nearby vertical clay wall facing northeast. In December 1971 and January 1972 no adult bees were found at these sites, although a few flaccid prepupae were found in cells. In the period May–July 1971, W. G. Eberhard (personal communication), in the course of other studies, noted activity of *Melitoma* around the nests, but does not know whether or not such activity began before or continued after his observations.

10. *Kourou, French Guiana.* The site was 2 km northwest of Kourou, on the first low sand ridge behind the beach (not over 3 m above sea level). The original forest had been partly replaced by mango and other introduced trees, but the area was wooded except for the largely open or brushy seaward slope of this sand ridge. *Ipomoea pes-capri* flowered on the sand near the high-tide limit, and other species were common in open or brushy areas.

In a tiny opening beside a fresh lagoon that contained water, except toward the end of the dry season, a house had been constructed; only the supports and the roof remained. The floor of the house had been made of sandy clay about 10 cm thick, possibly dried mud from the mangrove flats offshore, and about a dozen *Melitoma* nests had been made in an area of this flat floor about 10 x 25 cm. The deepest cells were in the bottom of the clay layer, but none extended into the sand beneath. Although the rest of the seemingly uniform floor contained no burrows of *Melitoma*, this small area was honeycombed with old burrows; presumably the site had been in use for several years. This is one of the few sites where *Melitoma* is known to nest in flat soil (see Table 1). Such a site must be acceptable here only because of protection from rain by the roof; mean annual rainfall is nearly 3000 mm.

11. *Quirigua, Guatemala.* Nests in a clay bank (Cockerell, 1913). The bees were perhaps active in February when other material was taken at this locality, but may have been identified from the dead dry specimens sometimes found in cells. [*segmentaria* or *marginella*]

12. *Tuxtla Gutiérrez, Chiapas, Mexico.* A nesting site near this locality was observed by R. B. Selander, from whose notes the following has been extracted: The site, 8.8 km by highway east of the city limits of Tuxtla Gutiérrez, was in the arid lower tropical subzone, altitude 440 m, along a cattle path between the Rio Grande de Chiapas and Highway 190, where the river is close to the highway. The nests were about 15 m from the lower end of a vertical clay bank along the path and overlooking the river to the south and southeast; the bank was over 20 m long, grading from 2 m high at the north end to 3 m high at the south end. The face of the bank was overhung in some places and was shaded quite effectively by vegetation growing on top of the bank and on the opposite side of the cattle path. During an entire afternoon at the site in July 1955, direct sunlight did not reach the face of the bank at any time.

The clay was fairly soft. Burrow entrances were crowded in many areas, filling the bank fairly uniformly, except for a zone between 0.6 and 1.5 m above the path which was marked with numerous gouges from cattle, which scraped their horns along the bank while being herded to and from the river. This habit of the cattle was an important factor in the ecology of this nest aggregation. In the worn central strip, many cells were exposed at the surface. Above and below, where the bank surface was relatively undisturbed, cells were from 2.5 to 10 cm deep.

When this area was visited in 1956, the site had been destroyed. A few burrows were found in other banks in the immediate vicinity, but no dense population as described by Selander.

In the town of Tuxtla Gutiérrez, however, burrows of *Melitoma* were observed in 1956 in the walls of several adobe houses near the river. In one large two-story house about 100 m from the river, 7 separate aggregations of a dozen or more burrows each were located just beneath the eaves. In the banks of the river only 3 cells of *Melitoma* were found. [*marginella*]

13. *Rio La Venta, Chiapas, Mexico.* At the crossing of the Rio La Venta by the Pan-American Highway, midway between Cintalapa and Ocozocoautla, at an altitude of 520 m, *Melitoma* nested in several banks and adobe buildings. On August 6, 1956, only two adults of *Melitoma* were seen flying. The various localities were as follows:

a. A west-facing bank along a dirt road which paralleled the river. Most of the bank was rocky and unsuitable for nesting, and only a few single burrows or small groups were found in scattered areas of soil. A major nesting site was located, however, near one end of the bank, in an area of only moderately hard soil. This portion of the bank was heavily eroded, but burrows were in a recessed area beneath an overhang of exposed roots.

b. A relatively new adobe house. Several hundred *Melitoma* burrows were near the top of the south-facing wall; they were protected by overhanging eaves. Only a few burrows were in the other three walls. An adjacent wattle-and-daub house had a few burrows near the eaves on the south side.

c. Vertical banks of an arroyo near the houses. A number of small to rather large aggregations of burrows were in the east- and west-facing earthen banks, 0.6 to 1.5 m high. The largest aggregations were in undercut portions of the banks, but only cells containing parasites (many exposed at the bank surface) could be found.

The owner of the houses pointed out other banks which he said had previously supported large populations which had been destroyed when the bank faces broke off during rainy weather. [*marginella*]

14. *Francia, Chiapas, Mexico.* At Francia, 12.9 km northeast of Cintalapa, altitude 810 m, *Melitoma* nested in a vertical, south-facing roadside bank about 2 m high and 15 m long. The nests were densest in a section 5 m long near the center of the bank, where the openings averaged about 2.5 cm apart. The bank was a sandy clay mixture and in April 1953 was dry and very hard. No adult bees were in flight. (Adults emerged in the laboratory from August 31 to October 15, primarily in September.) A small stream flowed a few meters from the site. A few *milpas* were nearby, and to the south the land was flat, extending off to the Rio Cintalapa. [*marginella*]

15. *Soyalo, Chiapas, Mexico.* Several nesting sites were examined near this town on the road between the Pan-American Highway and Simojovel, at an altitude of 1140 m. In 1953, more than 100 burrows were found in a small undercut roadside bank 6 km southeast of Soyalo (Fig. 1) near a stream. In 1956 none could be located in this bank or in the surrounding area, but scattered burrows were found in other banks near the Pan-American Highway. These banks were within 100 m of a stream but were largely too sandy to be suitable for *Melitoma* [*marginella*]

16. *Yautepec, Morelos, Mexico.* Bees were active on July 3, 1951, at an aggregation

of nests in a vertical clay bank of a temporary water course (Michener and Lange, 1958). [? *marginella*]

17. *Guanajuato, Guanajuato, Mexico. Melitoma* were found nesting in a number of adobe houses in or near the city of Guanajuato, altitude 2080 m. All such houses were near water sources, and some *Ipomoea* plants were in bloom near the water. On two visits—August 15, 1954, and August 16, 1956—a few adults of *Melitoma* were burrowing, but no large populations were found. A number of old burrows were also found in river banks a short distance west of the city. [? *marginella*]

18. *Fresnillo, Zacatecas, Mexico.* Nesting of *M. segmentaria* was observed 27 km northwest and 2.4 km southeast of Fresnillo, altitude 2230 m, in July and August of 1954 (Linsley, MacSwain, and Smith, 1956). In addition, a few old burrows were noted 14 km southeast of Fresnillo in 1954. Contrary to the situation at most other sites, there were no permanent streams and the bees were dependent, at least in part, on temporary waters remaining after the frequently torrential summer rains.

a. Many plants of *Ipomoea longifolia* bloomed at the locality 27 km northwest of Fresnillo, where a few burrows were found in the banks of a drainage culvert under the highway.

b. Extensive nesting was found in the banks of a deep wash 2.4 km southeast of Fresnillo. The wash was subjected to occasional floods, which on one occasion filled it completely; between storms a few small puddles remained in the bottom of the wash for several days. Aggregations of large numbers of burrows were in recessed sites along several hundred meters of vertical clay-like banks. A few adults were active and visited flowers of *I. longifolia* and *I. pringlei.* A discussion of these plants and their apoid visitors has been published by Linsley, MacSwain, and Smith (1956).

c. At the third locality, 14 km southeast of Fresnillo, *I. longifolia* was moderately abundant near a shallow reservoir, but the banks of a nearby wash were mostly too rocky for nesting. A few old burrows were noted in a pocket of clay-like soil in one bank.

19. *Concordia, Sinaloa, Mexico.* Old nests were found by G. E. Bohart in a wall of an adobe house 6.5 km west of Concordia (Torchio, 1974). Flight activity had only started in mid-September, and was intense in early October (G. E. Bohart, personal communication). [*marginella*]

20. *Culiacán, Sinaloa, Mexico.* A nesting site in the vertical wall of a dry arroyo, 18 km north of Culiacán (Parker, 1977). [? *marginella*]

Melitoma grisella

21. *La Junta, Colorado, U.S.A.* Nests were reported in banks (Cockerell, 1934a).

22. *Cheyenne County, Kansas, U.S.A.* A small aggregation of 4 nests plus deserted holes was evident in a bank (Hungerford and Williams, 1912).

23. *Atwood, Rawlins County, Kansas, U.S.A.* An isolated nest was found in a clay bank (Hungerford and Williams, 1912).

24. *Hot Springs, South Dakota, U.S.A.* Nests in large numbers in a bank (Cockerell and Porter, 1899, reported under the synonymous name *Entechnia dakotensis*).

Melitoma taurea

25. *Utica, Mississippi, U.S.A.* Ashmead's (1894) account of "*Epeolus donatus*" can only apply to *M. taurea.* Ashmead himself reached the same conclusion, according to

Robertson (1899); presumably the cleptoparasite, *Triepeolus donatus,* was investigating or entering the *Melitoma* nests and was mistaken for the nest-making bee. Several nests were in hard clay (presumably a horizontal surface) under a shed.

26. *Lebanon, Missouri, U.S.A.* The site was 16 km east of Lebanon, in a small south-facing bank along a roadway. The bank was nowhere over 0.3 m high, and was not vertical but only 20° to 50° from horizontal. Nests in it must therefore have been fully exposed to heavy rains in this mesic region, but about 30 nests were scattered along about 30 m of bank, and old nests indicated long occupancy of the site. These nests were in the most nearly horizontal exposed ground found by any of us to be used by *Melitoma,* although nests in level ground protected from rain are recorded at sites 10, 25, and 27. Those reported from level ground at sites 29 to 30 may or may not have been in protected locations. The Lebanon site was described by Michener (1975); on August 12 all the active bees were tattered females.

27. *Wicks, and elsewhere in St. Louis County, Missouri, U.S.A.* The site at Wicks described by Rau (1926, 1929) was in clay protected from rain under the porch of a house; most of the *Melitoma* nests were in horizontal soil, where the light intensity was low, above a clay bank. Small aggregations were in roadside banks or in soil brought up by trees uprooted in a storm. In the latter sites, where more or less vertical or overhanging soil surfaces were protected from rain by broken roots, aggregations seemed to thrive, but in the unprotected banks that Rau studied there were few nests and no continuity from year to year. As Rau (1934) noted, all sites were near water. The first flight activity was on July 12, the last on October 3, but the greatest activity appears to have been in mid-August.

28. *Keyes Summit, St. Louis County, Missouri, U.S.A.* Nests were in a roadside bank facing west, under protecting roots and vegetation (Rau, 1929).

29. *Wesco, Crawford County, Missouri, U.S.A.* Some nests were in banks, some in level ground, and three groups were in soil among roots of trees torn up by a tornado three years previously (Rau, 1929).

30. *Indiana* (localities not given), *U.S.A.* Nests are reported in banks or in flat, bare soil (Chandler, 1958) or in clay banks or earth retained among the roots of an overturned tree (Say, 1837; Ashmead, 1894).

DISCUSSION OF NEST SITES

There is no evidence that these four species of *Melitoma* differ significantly in the sites selected for nesting or in nest structure.

All sites studied by us (except perhaps 8) were near flowering *Ipomoea* plants, from which food resources are obtained, and near water, which is used to soften the hard clay in which nest burrows are usually made. The substrate was "fairly soft" at site 12 (Tuxtla Gutiérrez), but elsewhere its extreme hardness was commonly noted. A preference for hard soil was demonstrated at site 6 (Curitiba, Brazil), where *Melitoma* nests were restricted to local hard areas in extensive softer banks. Although sometimes isolated or scattered, nests are commonly in dense aggregations, sometimes of hundreds or thousands. The data often do not quantitatively distinguish such categories, but in Table 1 we have indicated sites at which scattered and densely aggregated nests have been reported.

Nests are usually in vertical clay banks, often most abundant beneath a root or other structure that causes an overhang (Fig. 1) or provides protection from weather. Sometimes they are in exposed sloping banks (20° to 50° from horizontal) as noted at site 26 by Michener (1975). A few sites (10, 25) were in horizontal clay, protected by a roof or other manmade structure. Ordinary unprotected horizontal ground is usually unsuitable for nesting by *Melitoma,* although used by other melitomine genera (*Ptilothrix, Diadasia*). There are, however, a few reports of nests of *Melitoma* in flat ground, not stated by the authors to be protected from the weather (some of those at sites 29 and 30).

Table 1 summarizes information on slope and degree of protection of sites. Although some populous nesting sites are in deep shadow, we believe that north-facing banks or walls are not or little used in the north temperate zone, and that south-facing banks or walls are not or little used in the south temperate zone. Banks with such exposures are clearly better illuminated than some of the deeply shadowed sites, but nevertheless appear to be avoided.

In primeval times, most nest sites were no doubt in banks along water courses. Only where streams have cut through clay, leaving vertical banks well above the water level, would good sites have been likely to occur. Human activities, however, have greatly increased potential nest sites, not only by the cutting of roadside and trailside banks, but by construction of adobe buildings, the walls of which make good nesting areas, especially where protected by overhanging eaves.

The extensive human use of adobe for construction has been primarily in the drier parts of Latin America. The same areas tend to have barrancas or washes with vertical

TABLE 1. Some characteristics of nest sites of *Melitoma segmentaria, M. marginella, M. grisella,* and *M. taurea.*

In dense aggregations	2, 4, 9(p), 10, 12(p), 13(p), 14, 15(p), 16, 18(p), 24, 27(p)
Scattered or isolated	1, 6, 7, 9(p), 12(p), 13(p), 15(p), 17, 18(p), 19, 23, 26, 27(p)
In natural, more-or-less vertical banks	1, 13(p), 14, 16, 17(p), 18(p), 20, 22, 23, 24, 27(p), 29(p), 30(p)
In man-made, more-or-less vertical banks	6, 9, 12(p), 13(p), 14, 15, 18(p), 27(p), 28
In man-made walls	2, 4, 5, 7, 8, 12(p), 13(p), 17(p), 19
In sloping banks	26
In horizontal ground	10, 25, 27(p), 29(p), 30(p)
In deep shadow	9, 25, 27(p)
Protected by overhanging roots, eaves, etc.	6(p), 7, 9(p), 10, 12(p), 13(p), 25, 27(p), 28
Unprotected, on vertical surfaces	1, 2, 3, 8, 9(p), 12(p), 13(p), 14, 15, 16, 17, 18, 27(p)
Unprotected, in sloping or horizontal soil	26, 29?(p), 30?(p)

Numbers refer to sites listed in the text; (p) = part, i.e., the statement applies to some but not all nests at the site.

walls in which the bees sometimes nest. By contrast, in the wet tropics adobe is usually impractical for construction purposes; and if erosion produces any vertical banks, they quickly become covered with dense vegetation. We have never seen such banks used by *Melitoma.* It is not surprising that the genus is rare in lowland wet tropical areas like the Amazon Valley. Extensive sandy areas also contain few nesting sites, as can be seen by the lack of *Melitoma* in most tropical beach communities where *Ipomoea pes-capri* is blooming abundantly. Yet at site 10, virtually on the beach in French Guiana, where clay had been made available by man close to the shore, *Melitoma* was present and visited *I. pes-capri* as well as other species of *Ipomoea.*

Several nesting sites observed in Mexico help to answer questions such as whether aggregations of this species are the result of (1) habitat preference or limited available habitats, (2) a tendency for progeny to return to the vicinity of the parental nest, or (3) a tendency of individuals to locate their burrows near others of their species.

Whereas in most localities available nesting sites were limited, in two areas suitable habitats appeared to be unlimited. At one locality near Fresnillo (site 18), several hundred meters of vertical clay banks over 2 m high bordered both margins of an intermittent stream. Here were several dozen aggregations of burrows whose positions might have been due to unknown differences in soil makeup or to irregularities of the bank surfaces, but appeared to be randomly distributed. More significant was the two-story adobe house in Tuxtla Gutiérrez (site 12), where 7 separate aggregations of a dozen or more burrows each were located in different places just beneath the eaves.

At site 2 in Entre Rios, Argentina, a small aggregation of nests was in an extensive, presumably uniform wall (Doello-Jurado, 1912), and at site 10 in Kourou, French Guiana, nests were in a small area of a much larger and uniform surface. Wherever appropriate observations could be made, old, empty burrows were in the aggregations among the occupied ones, but isolated burrows were often not near old nests. We interpret these findings as suggesting alternative (2), a tendency for progeny to locate their burrows near the parental burrow. We saw no evidence that an aggregation could arise in a single season, as occurs with some other bees, e.g., *Nomia melanderi*, for which alternative (3) appears to be correct (G. E. Bohart in Michener, 1974b).

We have repeatedly found old nesting sites, sometimes with hundreds of burrows, but without or almost without bees. Such sites have been seen or reported from the United States to Brazil. As *Ipomoea* and water are usually still present at such sites, we assume that extinction is probably due to parasites and predators. Our impression is that aggregations of *Melitoma* commonly die out after several years of activity. This is unlike the long survival of aggregations of some other solitary bees.

SEASONAL CYCLE

As indicated in Table 2, in both northern and southern temperate regions there appears to be but one generation of *Melitoma* per year. Activity is all in mid and late summer and early autumn (e.g., late February and March at site 6, Curitiba, Brazil—Michener and Lange, 1958; mid-July through September, a few moribund individuals in earliest October, at site 27, St. Louis County, Missouri—Rau, 1929).

In the tropical zone the situation is more complex. Bee specialists have not usually recorded activity or lack of it in any one tropical area over a long enough period of time to show what the seasonal cycle is. We have supplemented data obtained at nest sites (Table 2) with data from specimens taken by collectors (Table 3), but the result is information on the seasons when the bees do fly, with almost nothing on when they do not.

Nonetheless, long seasons of inactivity clearly occur in the marginal tropical areas. This is not at all surprising at localities at high altitudes, as in the state of Zacatecas, Mexico, where we have records of activity only from July through October. Cessation of activity in the Southern Hemisphere winter is probable in the states of São Paulo and Guanabara, Brazil, even though specimens have been taken in flight in both states in April (Table 3) when activity had already ceased at site 7 (Table 2). In the northern tropical parts of Mexico (e.g., the states of Sinaloa, Veracruz, Nayarit, and Jalisco), the season of activity appears to be considerably lengthened (Table 3).

In many more fully tropical countries, Table 3 indicates activity more or less throughout the year. This does not necessarily mean that flight occurs throughout the year in each area. Particularly in the cordilleran countries, there is such topographic and climatic diversity that different populations might well have different seasonal cycles. For example, at sites 12 to 15 in the state of Chiapas in southernmost Mexico, lack of adults was recorded in March, April, June, and July (Table 2), the populations existing at that time as immature stages only. Yet pinned specimens record capture of adult specimens in Chiapas state during the same months (Table 3). Less decisive examples are the following: At site 9 (Cali, Colombia) the bees were active during the period May to July 1971 (W. G. Eberhard, personal communication) but no adults were found the following December and January by C.D.M. At site 10 (Kourou, French Guiana) adults were active and taken on flowers in June and July (end of the wet season) but not during the rest of the year, in spite of year-round collecting. (As the population was small and localized, the flight season may have been longer, and at another locality in French Guiana adults were taken on flowers in March.)

TABLE 2. Months of flight activity by females of *Melitoma taurea, M. grisella, M. marginella,* and *M. segmentaria* at nest sites.

Site	Months of adult activity											
	Jan	Feb	Mar	Apr	May	Jun	Jul	Aug	Sept	Oct	Nov	Dec
1 (Buenos Aires, Argentina)		X	X									
2 (Entre Rios, Argentina)		X	X									
3 (Montevideo, Uruguay)			X									
5 (Puerto Alegre, Paraguay)			X									
6 (Paraná, Brazil)	-	X	X	-	-	-	-	-	-	-	-	-
7 (São Paulo, Brazil)	X			-								
8 (Lima, Peru)						X						
9 (Cali, Colombia)	-				X	X	X					-
10 (Kourou, Fr. Guiana)		-				X	X			-		
12 (Tuxtla Gutiérrez, Chiapas, Mexico)						-	-	X̲	X̲	X̲	X̲	
13 (Chiapas, Mexico)								X				
14 (Chiapas, Mexico)				-					X̲	X̲		
15 (Chiapas, Mexico)			-									
16 (Morelos, Mexico)							X					
17 (Guanajuato, Mexico)								X				
18 (Zacatecas, Mexico)							X	X				
19 (Sinaloa, Mexico)									X	X		
26 (Lebanon, Missouri)								X				
27 (Wicks, Missouri)						-	X	X	X	-		

Arabic numbers refer to sites listed in the text; sites 7 to 19 are in the tropical zone.
- = no activity; X = activity; X̲ = emergence of adults in the laboratory, but no field data.)

Such observations do not indicate, however, that *Melitoma* is everywhere strongly seasonal. Specimens from the small and relatively homogeneous wet tropical island of Trinidad have been taken in every month of the year except April and September. They have been collected at Curepe, Trinidad, in January, May, June, July, October, November, and December (material from the Commonwealth Institute of Biological Control). Adults may disappear in lower, drier places like Curepe for a short time in the dry season, but may also be continuously active. The number of generations per year in tropical areas is not known.

Although *Melitoma* is seasonal, at least in many areas, emergence is not well synchronized, even in temperate climates. Thus Rau (1929) found the first bees at site 27 (St. Louis County, Missouri), on July 12, but indicates that emergence was continuing at least as late as August 14.

Doello-Jurado (1912) indicates that males appear before the females and die before the last females. Adults of both sexes appear to be long-lived; there are no appropriate data on the subject, but Rau's (1929) account appears to assume that they survive for at least a month or two; he says "all summer." Mating occurs around the nest site.

TABLE 3. Months of flight activity within the tropical zone for *Melitoma segmentaria* (1) and *M. marginella* (2)*

	Jan	Feb	Mar	Apr	May	Jun	Jul	Aug	Sept	Oct	Nov	Dec
Bolivia and Paraguay	1	1	1						1	1		
São Paulo and Guanabara, Brazil	1	1		1						1	1	1
Paraiba, Brazil									1	1	1	
Pará, Brazil						1	1					
French Guiana			1			1	1					
Guyana and Surinam			1	1	1							1
Trinidad	1	1	1		1	1	1	1		1	1	1
Venezuela	1			1	1	1		1		1		
Peru	1	1	1	1	1	1	1			1	1	1
Colombia and Ecuador	1	1	1	1	1	1	1			1	1	1
Panama	1	1	1	1								1
Costa Rica			1				1	1	1			1
Guatemala to Honduras	2					2	2	1	2	2		2
Chiapas, Mexico		2	2	2		2	2	2				
Oaxaca, Mexico						2	2	1		1		
Veracruz, Mexico	2			2	2	2		2	2			2
Guerrero, Mexico							1	1	2	2		
Nayarit and Jalisco, Mexico			2				2	2				2
Tamaúlipas and San Luis Potosí, Mexico						2	2		2			2
Sinaloa, Mexico			2					1, 2	2	2		2
Zacatecas, Mexico							1	1	2	2		

*Based on label data from pinned museum specimens assembled by Dr. Thomas J. Zavortink.

Rau found numerous males attracted to a burrow which presumably contained an emerging female, and to a callow female that he had removed from its cell. He also reports a male mating with an obviously older female that was making a nest. We have never seen mating on flowers or elsewhere, but have observed males flying around the nesting sites as well as at flowers of *Ipomoea*.

Published data on the durations of immature stages are lacking. It is clear that larvae develop moderately rapidly, so that larval feeding is complete a few weeks after egg laying. The larva then remains dormant as a prepupa until the approach of the following flight season. Michener (1975) found such prepupae at site 26 (Lebanon, Missouri) in mid-August. Since the oldest larvae produced that season were still rather small, he suggested that for some prepupae, development may be arrested until a second year. In large samples of cells from Mexico brought into the laboratory (and thus exposed to abnormal conditions), no emergences were delayed until a second year. This finding does not necessarily invalidate the suggestion, however. When the prepupal dormancy is terminated, pupation occurs. Under laboratory conditions, the pupal period for 6 males from Mexico was 26 to 30 days, with a mode at 27 days; for 13 females, 27 to 30 days, with a mode at 28 and 29 days.

NESTS

Nest construction and nest structure have been described previously, in most detail by Rau (1929) and Michener and Lange (1958). The present account summarizes these papers and adds certain details and information on variations.

The nest burrows are 5-6.5 mm in diameter. When in vertical surfaces, they are more or less horizontal or slope slightly upward or occasionally downward. When in horizontal surfaces (as were all at site 10 and about 90% of the nests in the protected situation at site 27), the burrows extend approximately vertically, downward. Some females re-use old burrows, but in such cases commonly construct new tunnels off of them. To dig in the usually extremely hard, dry clay, a bee brings water to the site in her crop, regurgitates it—thus softening the clay—and then excavates with her mandibles. Water is usually obtained at the edges of ponds or streams, but the bees do not usually alight on the water surface as does the related genus *Ptilothrix*.

Pellets of moist clay removed from the burrows are used to form an entrance tube, which lies on the substrate (contrary to an incorrect statement that these tubes project —Michener, 1975). On vertical surfaces, such tubes ordinarily extend downward from the burrow entrance, following contours or irregularities of the surface; but when crowded, the tubes may be directed laterally or rarely upward. Entrance tubes of nests on horizontal surfaces lie flat or slightly inclined on the substrate. The tubes are relatively smooth on the inside and rough on the outside, because of the way in which the bees manipulate the mud. Early in nest construction, as the burrow is deepened, the bee loosens moistened soil, carries it held against the body with the forelegs as she backs to the mouth of the burrow or entrance tube, and then passes the pellet of soil to the rear and pushes it into place and shapes it with the pygidial part of the abdomen, pressing and smoothing the inside but not touching the outside of the entrance tube.

Usually the entrance tube is constructed with a median longitudinal fissure in addition to the terminal opening, as illustrated in Figure 1 (site 18) and by Rau (1929, Figs. 9 and 13) for nests at site 27. Some fully formed tubes lack such a fissure, however, as in the one from Kansas illustrated by Hungerford and Williams (1912). Entrance tubes are commonly 2 to 3 cm long. Some burrows, however, entirely lack entrance tubes or—more often—have a mere collar of dried mud around the entrance. While this lack may be due to the failure of some bees to construct entrance tubes, it often results from destruction by weather, animals, or man. As the tube is built early in burrow construction, after which excavated substrate is dumped in the form of loose pellets, it is not surprising that destruction of the tube of an older nest does not result in its reconstruction. The bee goes on with its activities, constructing and provisioning cells, going in and out of the unadorned nest entrance with no evident difficulty or hesitation. The bees at site 10 (Kourou, French Guiana) made less obvious entrance tubes than any other population studied. Most entrances had only collars; the longest tube was 1.5 cm long and lacked a fissure.

The female constructs several branch burrows, with up to 4 cells in linear series in each branch (see Figs. 8 and 9 in Michener and Lange, 1958). Occasionally a branch contains only a single cell, or the branch may be only as long as one cell, so that a cell is constructed projecting at an angle to those in a series. The tendency is for the branch burrows (including the inner part of the main burrow, if such a thing can be recognized) to become almost completely filled with cells, while the main burrow leading 1-6 cm from the entrance to the first cells is often left empty and open to the outside, even after all cells are complete. While this is the situation as described by Rau (1929) and Michener and Lange (1958), and as observed by C.D.M. in Missouri, French Guiana, and Brazil, E.G.L. and J.W.MacS. found in Mexico that in most cases the female eventually plugs a portion of the burrow with loose mud pellets and fashions a concave, smoothed, surface plug. A portion of the turret may be used in this construction.

Where the nests are dense, the burrows (and cells) from various years of occupancy are so interwoven that the structure of any one nest is difficult to ascertain. While some nests have a dozen or more cells presumably made by a single bee, many (Michener and Lange, 1958, Fig. 8; Michener, 1975, Fig. 1) have only 2, 3, or 4 cells. As implied by Rau (1926), many females probably make more than one nest.

Each urn-shaped cell (Fig. 2, and Michener and Lange, 1958; Michener, 1975) occupies a segment of burrow that is enlarged and shaped to accommodate the cell. Cellular orientation varies from horizontal to vertical, with the entrance up; usually the cells are inclined.

Each cell is constructed from soil (mud) within the burrow and can easily be removed from the substrate intact, except at site 10 (Kourou, French Guiana; Fig. 2, center left), where most but not all adhered to the substrate. Cells are rather variable in size, even at a given site. For most populations, cell entrance diameters are 5.0-5.5 mm; maximum inside diameters, 7.0-7.5 mm; maximum outside diameters, 9.5-10 mm; and outside length, 14-16 mm. At site 10 the bees tended to be small, and comparable figures are 4.5-5.0, 6.5-7.5, 8.5-10, and 12-13 mm. The maximum measurement of outside diameter (10 mm) was unusual, being based on a cell with walls about 2 mm thick. Most cells at the same site had walls about 1 mm thick, although there were curious irregularities, like a cell with a cap 2 mm thick instead of about 0.6 mm, as was usual at this site. While the cell wall is usually against the wall of the roughed-out cavity in which it is constructed, part of the wall can be built up free-standing, as where a cell is constructed partly in a previous season's cell or in an old cell of an associated species like *Ancyloscelis apiformis.* The outer surface of the cell is rough but the inner surface is very smooth (Fig. 2) and exceedingly thinly coated with a smooth, dark, wax-like secretion which reaches (site 10) or almost reaches (site 6, Curitiba, Brazil—Michener and Lange, 1958) the cell closure.

After the construction of a cell, the female provisions it with a mass of *Ipomoea* pollen, to which some nectar is added, the whole filling the basal third of the cell. The volume of the provisions (as well as the size of the cell) varies widely, and on drying the provisions may be removed like a ball of pollen. However, the fresh provisions conform to the basal walls of the cell, the upper surface usually convex or with a central elevation. In some cells the amount of liquid placed with the pollen causes some flow of the provisions if the cell is disturbed. After provisioning, an egg is placed beneath and to one side of the center of the food mass. In a few cases, where the food is incompletely packed and an air space remains in the side of the provisions, the egg is found within this cavity.

After oviposition, the female closes the cell with a mud cap 0.6 to 2 mm in thickness. The cap has a strong spiral pattern on the inner surface (Fig. 2), which has no wax-like lining. The outer surface of the cap is usually beautifully smooth and concave, and forms a concavity which receives the base of the next cell in series. At site 10, in some cells the outer surface of the closure was not smooth; in all cases such cells were not followed by another in the series. The females may have died before completing such caps, or perhaps in this population caps are smoothed only if another cell is to be constructed.

Within each cell, the egg hatches and the young larva eats its way up and around the pollen mass. This larval feeding does not deeply channel the pollen mass, as is the case with some *Ptilothrix* and *Diadasia* species. Before the completion of feeding, the larva begins to deposit a thin layer of feces over the inner surfaces of the cell. At site 6, dark feces were pressed against the cap; then the whole inner surface of the cell, including the dark feces, was covered with a thin layer (0.3–0.4 mm thick) of pale feces consisting of pollen shells. At sites 10 and 26 all the feces were pale, except that in a few cells at site 10 there were scattered dark feces on the inner surface near the cell cap. Although individual larval feces are often unrecognizable, at site 10 the light feces were more or less individually distinguishable, oriented longitudinal to the cell, then crushed, no doubt by the body of the larva, so that they formed a thin layer.

After the completion of feeding, a thin application of silk is usually made upon the fecal layer. A light brown cocoon, so thin as to be inconspicuous, is thus formed, consisting not only of silk fibers but of the usual amorphous matrix. At site 6 the cocoon was noted as incomplete (i.e., absent) in the bottom of the cell, and thickest near the cell cap. At site 10 the cocoon was sometimes complete, but was in other cases absent, not being recognizable even with a binocular microscope in one cell containing a pupa. At site 26 (Lebanon, Missouri) cocoons were complete. Thus much variation exists in the extent of the cocoon.

The last-stage larva after defecation (i.e., the prepupa) remains dormant until pupation in the following season, or possibly until subsequent years as suggested by Michener (1975).

The above details of nesting behavior correspond in general to the generic and tribal attributes described by Linsley, MacSwain, and Smith (1956) and Michener and Lange (1958). Variations in cocoons and in distinctness of larval feces would require minor modifications of tribal or generic charcterizations given in these accounts. So also would the use of horizontal nesting sites by some aggregations.

COMPETITORS ON FLOWERS

Our observations of other insect visitors to *Ipomoea* flowers were mostly made at site 18 near Fresnillo, Mexico. In that area a number of insect species were competing with *Melitoma segmentaria* for the pollen of *Ipomoea.* The more significant of these were the several kinds of lepidopterous larvae which were abundant in mid-July; another emphorine bee, *Ptilothrix sumichrasti*; a scarab beetle, *Euphoria basalis*; and a blister beetle, *Lytta variabilis.*

The scarab beetle was abundant in all areas near Fresnillo where *Melitoma* was studied. It is a common cetonid in central Mexico; Casey (1915) states that it is abundant from Durango to Mexico City, and Sallé (1833) reports its association with squash flowers. We, too, found this beetle in *Cucurbita* flowers at several localities, but not as abundantly as in the blossoms of *Ipomoea,* where up to 6 individuals occurred in a single flower. Shortly after the opening of new flowers each morning, the scarab beetles enter them, remain eating the floral parts throughout the day, and spend the night within the wilted blossoms. On clear mornings the bees undoubtedly gather their pollen before the beetles transfer to the new flowers, but on overcast or rainy mornings the beetles may invade flowers and destroy the pollen before the bees can gather it. Only half the flowers in the survey area produced fruits, perhaps because of the activity of *Euphoria* and other less common insects of similar habit.

Ptilothrix sumichrasti was the only other bee species consistently observed in *Ipomoea* flowers near Fresnillo, and unquestionably it competes with *M. segmentaria* for pollen. It has a much narrower distribution and was not found in all areas where the latter species occurred. Like *Melitoma,* it gathers pollen only from *Ipomoea,* but we found it associated mainly with *I. longifolia* and *I. pringlei,* whereas *Melitoma* visits a variety of species of the genus *Ipomoea.* In contrast to *Melitoma, P. sumichrasti* nests in flat ground closer to sources of water and pollen. The relative abundance of these two species is apparently controlled by the availability of suitable nesting sites. Wherever adequate sites for both species occurred, *Melitoma* was the dominant species. At a locality 27 km north of Fresnillo (site 18), the only bank was a small one formed by the construction of a drainage culvert, although there were extensive areas of flat or gently sloping bare ground. Here the proportion of the two species was about 10 to 1 in favor of *Ptilothrix.* The opposite was true 2.4 km south of Fresnillo, where there were both extensive banks on each side of a stream bed and large areas of flat barren ground.

Ancyloscelis apiformis is another bee that is an oligolectic visitor of *Ipomoea* flowers and must compete with *Melitoma.* We have observed the two species at flowers on the same plant at sites 9 and 10. *Ancyloscelis* may have a longer period of activity in the tropics than the strongly seasonal *Melitoma*.

Other bees also visit *Ipomoea* flowers. We have no quantitative data to offer on the subject, but *Trigona,* halictids, and males of the curcurbit oligoleges *Peponapis* and *Xenoglossa* are often common on *Ipomoea* in Mexico and Central America. In the eastern United States, *Cemolobus* is an oligolectic eucerine bee genus on *Ipomoea.*

As noted in the following section, *Nemognatha chrysomeloides* is a beetle whose larvae parasitize *Melitoma.* The adults are probably competitors of *Melitoma,* for they frequent the flowers of *Ipomoea,* from which they suck nectar with their long mouthparts (Fig. 3, right), like the bees.

ORGANISMS ASSOCIATED WITH *MELITOMA* CELLS IN MEXICO

More than 1900 intact cells of *M. segmentaria* and *M. marginella* were collected from 5 nesting sites in central and southern Mexico and held by E.G.L. and J.W.MacS. in the laboratory to determine their contents. The results are presented in Table 4. These data show major differences, both in kinds and in frequencies of organisms associated with *Melitoma* in different localities. We have no evidence that the parasites and predators distinguish between the two species of *Melitoma.* Our belief is that the differences shown in Table 4 between associates of *M. segmentaria* (site 18) and of *M. marginella* (sites 12–15) is due to environmental factors (semidesert plateau vs. lowland tropics) rather than the difference in species of *Melitoma.* The differences among the 4 *marginella* sites in the state of Chiapas are discussed below.

Pyrota

A few coarctate larvae of an unidentified meloid were found among the cells of *Melitoma* at site 14 (Francia). No such larvae were found in the cells; they must leave the cells prior to pupation. It was therefore impossible to determine the percentage of parasitism. In the sample of 800 intact cells, dead (unsuccessful) larvae of *Pyrota* were found in 0.3%, and we believe these to be the species represented by the coarctate larvae outside the cells. At site 13 (Rio La Venta) no resting larvae were found, but 30 cells, or 6.7% of the total, contained dead first- through third-stage larvae. Identifications were based on the characters of the primary larvae. The scarcity of fully fed resting larvae may indicate that the *Pyrota* is not usually a successful parasite of *Melitoma.*

Meloe laevis

This species occurs from the central United States through Central America. Dead larvae in various stages of development (perhaps killed by high temperatures during transportation) were found in 7 cells from Soyalo. Although no adults were found, there is no reason to assume that the species does not develop at the expense of *Melitoma.* Distributional data clearly indicate, however, that other bee species must also serve as hosts. Identification was based upon the characters of the primary larva.

TABLE 4. Percentages of cells from central and southern Mexico containing *Melitoma marginella* (sites 12–15), *M. segmentaria* (site 18), and various parasites or other associates

Sites	12 (Tuxtla Gutiérrez)	13 (Rio La Venta)	14 (Francia)	15 (Soyalo)	18 (Fresnillo)
Number of cells:	302	447	800	63	300
Date of collection:	June 22, 1955	Aug. 6, 1956	April 3, 1953	March 1, 1953	Aug. 6, 1956
Date examined:	Dec. 1955	Aug. 1958	Aug.- Oct. 1953	Aug. 1953	Nov. 1958
Melitoma	46.0	8.1	30.5	30.2	92.7
Pyrota sp.	—	6.7	0.3	—	—
Meloe laevis	—	—	—	11.1	—
Tetraonyx	—	—	0.1	—	—
Nemognatha chrysomeloides	12.6	44.1	17.0	28.6	—
Plega melitomae	—	7.2	7.8	—	—
Dasymutilla canina	3.6	0.2	0.4	—	—
Monodontomerus sp.	—	—	—	7.9	—
Monodontomerus mexicanus	—	—	—	—	0.3
Anthrax cintalapa	1.7	0.9	1.5	11.1	—
Trogoderma sp.	—	0.2	—	—	—
Chaetodactylus sp.	—	—	2.4	3.2	—
Mold-infested cells:	12.6	32.7	40.4	7.9	7.0
Pollen balls (no egg):	16.2	—	—	—	—
Miscellaneous and unidentifiable:	7.3	—	—	—	—

Tetraonyx

A *Melitoma* cell containing a dead depauperate adult of a species of *Tetraonyx* was broken into while making the collections at site 14 (Francia). It is doubtful if this meloid is a regular parasite of *Melitoma.*

Nemognatha chrysomeloides (Figure 3, left)

This meloid beetle occurs in populations of *M. marginella* in Mexico, and thence south through most of the range of *M. segmentaria* to Argentina. It was recorded as a parasite of *Melitoma* by Linsley and MacSwain (1957) and Parker (1977). The species was represented in samples from sites 13, 14, and 15 in Chiapas and site 20 in

Sinaloa, but we did not encounter it at sites 17 or 18 (Guanajuato and Fresnillo). Cells from site 15 collected on March 1, 1953, contained only coarctate larvae of the *Nemognatha.* Samples taken from site 14 at the end of July contained larvae and pupae; at the middle of September, larvae, pupae, and adults; and the same in mid-October. At site 12 (Tuxtla Gutiérrez), the species was in the late-feeding grub phase on June 22, and adults emerged from October 14 through December 25 in the laboratory.

The adult beetles frequent the flowers of *Ipomoea,* upon which they presumably oviposit, and have excessively long mouthparts, like those of their host bee, suitable for obtaining nectar from the depths of the flower (see Fig. 3, left). Our observations suggest that the life history of this species is similar to that reported for other species of the genus (Linsley and MacSwain, 1952a). The association of the adult beetles with the only pollen source of the bee probably explains the high percentages of parasitism. We have excavated some nest aggregations that were inactive and in which almost every cell contained evidence of this beetle, indiating that it may be a primary factor in the extinction of aggregations.

Cymatodera hopei
(Figure 4)

This clerid, originally described from Mexico, has been recorded by Gorham (1882) from Orizaba, Parada, Guanajuato, and Oaxaca. Horn (1880) redescribed it from Texas (under the name *gigantea*), and Corporaal (1950) listed it from California.

Larvae of *C. hopei* were taken from adobe walls being utilized by *Melitoma* at site 17 (Guanajuato). They exhibited a remarkable ability to tunnel through walls so hard that a mattock and chisel were required to extract the bricks. In what was probably searching for larvae and pupae of *Melitoma,* they completely traversed the walls from one side to another (1 m).

The larvae are apparently long-lived and extremely voracious. A nearly mature individual captured on August 15, 1954, consumed 8 bee larvae before pupating. The pupal period required approximately one month.

Plega melitomae
(Figures 5, 6)

The mantispid *Plega melitomae* was described from specimens reared from cells of *Melitoma marginella* from site 14 (Francia—Linsley and MacSwain, 1955). The cells analyzed for contents (Table 4) yielded 61 (7.8%) which contained mantispids in various stages. A similar infestation was found in the sample from site 13 (Rio La Venta) where 22 cells (7.2%) contained various stages of *Plega.* Only one example was taken by Selander at site 12 (Tuxtla Gutiérrez). Most samples contained only dead first-instar larvae or mature larvae within their cocoons. In one place at site 13 on August 6, 1956, several cast pupal skins and one living adult female were taken on the surface of the undercut bank from which the principal cell sample was collected. A number of active first-instar larvae were found in cells that had just been provisioned.

The presence of large numbers of dead first-instar larvae in some cells (in one case, 27), the shed skins of first-instar larvae in cells in which the *Plega* matured, and the absence of egg shells, indicates that entrance is made while the cell is being provisioned. However, whether the female oviposits in or near the burrow entrance (probable), or in a site where the larvae can attach themeselves to the bees and be carried to the cell (unlikely), or whether some entirely different method of entering is utilized, remains to be determined. The presence of a female on the bank at site 13 favors the first alternative.

As nearly as we can reconstruct the life history of *P. melitomae,* it is as follows: The primary larva enters the cell during provisioning and then waits until the development of the bee larva is complete, probably attaching itself to the surface of the host or remaining closely associated with it. After the host larva has completed defecation and cocoon formation, the *Plega* larva attacks it, feeds and develops rapidly, and then constructs its own cocoon within that of its erstwhile host. In this cocoon it overwinters as a larva. After completion of further development, the active pupa chews its way out of the cocoon and, moistening the cell wall with an oral secretion, emerges from the cell and crawls to a suitable place for the adult to eclose. The pupa usually attaches itself to a vertical surface, the skin splits, and the adult wriggles out. A larva taken by Selander at site 12 in June pupated sometime between August 1 and August 10. Among 22 adults from site 14 (Francia) that emerged in the laboratory, the earliest appeared between June 16 and June 26, presumably close to the latter date. The latest emerged between July 26 and August 1, the total span of emergence in our samples slightly exceeding a month. Adults emerged from samples taken at site 13 from August 18 to September 1 and later.

Dasymutilla canina
(Figure 7)

Adults of both sexes of this mutillid were taken about site 14 (Francia) in March and April. Cell samples in September and October contained dead pupae and incompletely transformed dead adults. However, a number of adults had emerged before the above samples were taken, and the percentage of parasitism was undoubtedly higher than is indicated in Table 4. The species was also present at site 12 (Tuxtla Gutiérrez), adults emerging from August 1 through August 10. At site 15 (Soyalo), larvae were found in 18 out of 63 cells examined.

Melitoma is probably only one of a number of hosts of *Dasymutilla canina.* Mickel (1928a) has recorded *D. fulvohirta* as a parasite of *Anthophora occidentalis,* and *D. foxi* as a parasite of *Diadasia* sp. Linsley and MacSwain (1942a) reared *D. aureola* from cells of *Anthophora bomboides stanfordiana,* and Hurd (1951) has reported the association of *D. aureola* with *Melissodes robustior* under circumstances suggesting a host-parasite relationship. Fattig (1943) reported *D. asopus bexar* (as *harmonia*) from cells of *Anthophora abrupta* and *A. ursina.* MacSwain (1958) recorded *D. dugesii* from the cells of *A. occidentalis*. In any event, *D. canina* is at least the sixth *Dasymutilla* species known to be parasitic on American anthophorid bees. Others are recorded from *Bombus, Epibembix, Microbembix, Polistes, Myzinum,* and *Trypoxylon,* and suspected as parasites of *Halictus, Dianthidium, Philanthus, Cerceris,* and *Sceliphron* (Krombein, 1951), and *Nomia* (Hurd, 1951).

Monodontomerus

Five of the cells among the 63-cell sample from site 15 (Soyalo) contained chalcidoids which were presumed to be *Monodontomerus*. At site 18 (2.4 km south of Fresnillo) several *Melitoma* cells were excavated which contained dead adults of both sexes of *M. mexicanus*.

Gahan (1941) has recorded *M. mandibularis* as a parasite of *Melitoma taurea*, and Rau (1947) has described some of its habits as a parasite of *Anthophora abrupta* in Missouri. Linsley and MacSwain (1942b) have also provided an account of *M. montivagus* as a parasite of *A. linsleyi* in California. MacSwain (1958) noted *M. montivagus* from the cells of *A. occidentalis*, and *M. mexicanus* on adobe walls which contained large populations of *A. marginata*.

Anthrax cintalapa
(Figure 8, left)

This species was described from material reared from cells collected at site 14 (Francia—Cole, 1957). It was also present at sites 12, 13, 15, and 20, and is apparently regularly associated with *Melitoma*, although percentages of parasitism were not high in any of the areas, and the fly is also parasitic on unrelated bees (Megachilidae) (Marston, 1970). Emergence in the laboratory was erratic and spread over much of the year. If we have correctly identified the greenish eggs plastered in large numbers on the banks around the burrow entrances, this species has oviposition habits unlike any encountered previously in this family. Samples of these eggs were placed in a container, and numbers of first-instar bombyliid larvae were present several weeks later.

Bertoni (1929) and Parker (1977) report other *Anthrax* species from cells of *Melitoma* at sites 5 and 20.

Anthrax mexicana
(Figure 8, right)

This species was taken on the wing at sites 13 and 14. Adults emerged from cells of *Melitoma* that had been appropriated by *Centria trigonoides*. It is not evident whether or not it is also a parasite of *Melitoma*.

Trogoderma

One cell at site 13 (Rio La Venta) contained shed larval skins of the dermestid genus *Trogoderma*. The cell contents had been completely destroyed by the beetle larva, which had escaped through a small hole cut in the side of the cell. It is likely that this same hole provided an entrance for the larva. A number of dermestids have been associated previously with cells of bees (Linsley and MacSwain, 1942b; Beal, 1954).

Chaetodactylid Mites

At site 14 (Francia), 2.4% of the cells examined contained hypopi of *Chaetodactylus* species near *ludwigi*. Most of these cells contained large numbers (usually hundreds) of mites, which appeared to have consumed all or substantial amounts of the pollen. Examination of the contents of infested cells failed to reveal fragments of bee eggs or larvae. This fact suggests that the egg or young larva is also destroyed by the mites, or that the latter develop successfully only in cells in which oviposition did not take place.

Since the hypopus stage is dependent upon phoretic transport from the cell in which it develops to a new cell where further development and reproduction occur, burrows of bees which arrange their cells in a series would seem to be suited to these mites. Bees emerging in cells lower down in the series would readily become contaminated if they pass through mite-infested cells. This would account for the frequent association of this group of mites with *Osmia, Xylocopa,* certain anthophorids, etc., which arrange their cells serially.

Microorganisms

The following report by the late E. A. Steinhaus on the microorganisms associated with cells of *Melitoma marginella,* from site 14 (Francia), is quoted in full because of its informative nature.

"The results of our [microbiological and pathological] examination have convinced us that the problem is of much greater magnitude than we originally supposed and that ordinary diagnostic procedures are not, in themselves, likely to give a complete or clear-cut picture of the role of microorganisms in the ecology of the bee concerned."

The fungus-infected larvae and cells were divided into five groups, as follows:

"(1) Yellow to brown collapsed larvae with no external evidence of fungi. (Upon cultivation, these almost invariably yielded blue and green fungi; no brown fungi were isolated from any of these specimens. The blue and green fungi were *Aspergillus versicolor* (Vuill.) Tiraboschi. (This identification was confirmed by Dr. Paul L. Lentz, Associate Mycologist, Bureau of Plant Industry.)

"(2) Dry gray larvae, occasionally with patches of flimsy mycelial growth, and occasionally with a peculiar encrusted raised growth containing conidia. (Many of these specimens contained what appeared to be a Microsporidia—Class Sporozoa, Subclass Cnidosporidia—identified as probably belonging to the genus *Plistophora.* Certainty as to the identification of this organism was not possible, since if it is a protozoan, vegetative stages were not available. The average size of the elliptically shaped spore-like bodies was approximately 2.6 by 5.0 microns. They occurred in groups of varying number, usually more than 16, a fact that characterizes the genus *Plistophora.*)

"(3) Cells, the contents of which were covered with brown fungi. (These fungi were difficult to identify as to species, but belong to the *Aspergillus flavus-oryzae* group. In attempting to make a specific identification of the brown species, Dr. Lentz says: 'It has spiny conidiophores, sterigmata in both single and double series, sclerotia which are white when young and blackish when older, conidia echinulate, and is 6–7 microns in diameter.')

"(4) Cells, the contents of which were covered with yellowish fungi. (These were also *Aspergillus*, probably belonging to the *A. flavus-oryzae* group.)

"(5) Cells, the contents of which were covered with blue fungi. (This appears also to be a strain of *Aspergillus versicolor* (Vuill.) Tiraboschi. These blue fungi appeared to have neither perithocia nor sclerotia. At first we thought these blue strains belonged to the *A. nidulans* group, and since they produced an abundance of Hulle cells we thought perhaps they might be *A. caespitosus* Raper and Thom. However, Dr. Lentz felt they are closer to *A. versicolor*, since their cultural characters appeared to fit those of this species very well. Hulle cells of the *A. nidulans* type are occasionally seen in *A. versicolor*. Members of the *A. versicolor* group are cosmopolitan in distribution. They occur regularly in the soil, upon decaying vegetation, and on many products exposed to occasional moist air or undergoing slow decomposition.)

"Some cells had a mixture of fungi, and in other cases the fungi seen appeared to be saprophytes or purely adventitious species. It would not always be safe, therefore, to rely on color to gain a provisional identification of the fungus concerned. Also to be remembered is the fact that fungi exhibit a great deal of variation with regard to color, the same species showing marked differences in color under various conditions and circumstances. On a few occasions *Beauveria bassiana* (Balsamo) Vuill., or something close to it, was isolated from the material in these accessions."

ASSOCIATES OF AGGREGATING ANTHOPHORIDAE IN NORTH AMERICA: COMPARATIVE COMMENTS

Our data on organisms associated with cells of *Melitoma segmentaria* and *M. marginella* permit comparisons with similar data on some other North American anthophorid bees which occur in large aggregations (Table 5). The records for *Anthophora occidentalis* are from Mickel (1928b), Porter (1951), MacSwain (1958), Hobbs, Nummi, and Virostek (1961), and Ferguson (1962). Each presents the number of infested and uninfested cells, and the data are from widely separated localities (Colorado, Texas, New Mexico, Idaho, and Alberta). Information on *A. linsleyi* is from Linsley and MacSwain (1942b); on *A. flexipes* from Torchio and Youssef (1968); on *A. edwardsii* from Thorp (1969); and on *A. peritomae* from Torchio (1971). Data on *Diadasia consociata* and *D. bituberculata* are from papers by Linsley, MacSwain, and Smith (1952) and Linsley and MacSwain (1952b). Similar information on *A. bomboides stanfordiana* and *D. vallicola* are available in Linsley and MacSwain (1942a) and Ferguson (1962). Other authors have presented records of various insects associated with the aggregated burrows and cells of anthophorid bees, but since they do not give numerical data, and since space does not permit listing of all such data here, their results have not been considered below, except as noted. The following remarks are presented as possible explanations of the associations shown in Table 5.

Adults of the meloid genus *Lytta* feed on leaves, flowers, or both. North American species for which bee hosts are known feed on flowers; Selander (1960) states that "There is evidence that some species of *Lytta* feed on the same kinds of plants that their bee hosts utilize as pollen sources." Following feeding and copulation, females deposit masses of eggs in the soil in the same habitat. Larvae then search over the ground for the burrows of solitary bees and, on entering, consume the contents of one or more cells. Where aggregations of bees' nests occur near the feeding and oviposition site of a *Lytta* species, they may be preyed upon. This would explain why some populations of these bees have been reported to support as many as three species of *Lytta* with more than 6% of the cells destroyed, while in other populations of the same bee species *Lytta* is absent. A similar pattern is indicated for the genus *Pyrota*.

Adults of *Meloe* appear before their host bees and, since they are flightless, remain near their point of emergence, where they feed on leaves or other vegetation. Eggs are deposited in the soil, and the first-instar larvae hatch at about the time of adult bee

TABLE 5. Insects associated with aggregations of some anthophorid bees in North America.

Associated Group	*A. edwardsii*	*A. flexipes*	*A. linsleyi*	*A. occidentalis*	*A. peritomae*	*Melitoma* spp.*	*D. bituberculata*	*D. consociata*
	Bee species							
Meloidae								
Lytta	+		+				+	
Pyrota						+		
Meloe	+					+		
Tricrania	+			+				
Nemognatha	+		+	+		+		
Hornia			+	+				
Tetraonyx						+		
Zonitis		+						
Rhipiphoridae							+	+
Bombyliidae	+	+	+	+	+	+	+	+
Mutillidae	+		+	+		+	+	
Mantispidae						+		
Chalcidoidea			+	+		+		
Parasitic bees	+	+	+	+	+			
Scavengers	+		+	+		+		

A. = *Anthophora*; D. = *Diadasia*.
**M. segmentaria* and *marginella*.

activity. These larvae climb vegetation, and when contacted by a bee, cling to its hairs. Those which attach to females of the right species gain entrance to the new burrows and, after leaving the bee, consume the contents of one or more cells. The hazards involved in attaining the host by the larva may explain both the relative rarity of *Meloe* and the fact that most populations of known host species are not infested. Of the bee genera treated here, *Meloe* has been associated with several species of *Anthophora*, one species of *Melitoma*, and no species of *Diadasia.* The circumstances which limited it to only one of the four major populations of *N. marginella* studied by us are not clear.

Adults of *Tricrania* are not known to take food. The two species whose habits have been studied oviposit under loose objects near the burrows from which they emerge. One of these species is flightless, and is associated with bees of the genus *Colletes*, whose burrows occur in aggregations. The other species is rarely taken in flight, and its known hosts are a number of species of megachilid bees and two species of *Anthophora.* As the first-stage larvae are phoretic on adult bees, they may reach host cells in the same manner as *Meloe*.

Species of *Nemognatha* have been recorded in small numbers from the cells of *Anthophora*, in large numbers from those of *Melitoma marginella*, and are unknown from any species of *Diadasia*. Adults of each species show preferences for flowers of particular species or genera, and females deposit masses of eggs on the undersides of buds. The meloid larvae hatch in close synchrony with the opening of the flowers and arrange themselves on the flower parts. When the flower is visited by an insect, the larvae cling to its body hairs with their mandibles. E.G.L. and J.W.MacS. found many larvae of a species which oviposits on *Cirsium* attaching themselves to unsuitable hosts such as Lepidoptera and Diptera. Those attaching to females of an appropriate bee species are carried to the burrows and leave the bee in a cell being provisioned. Here they eat the bee egg and then the pollen and nectar. Since *Anthophora* species visit a variety of plants for nectar and pollen, it is not surprising that there is a relatively low percentage of parasitism. *Melitoma* and *Nemognatha chrysomeloides*, however, are largely restricted to the same plant; there are few other insect visitors to it, and it follows that the percentage of parasitism may be high. There is nothing to suggest that aggregation of nests in bees of these two genera favors these meloids. From a knowledge of the flowers utilized by *Diadasia* and by *Nemognatha*, it is clear that the two groups would not come into contact except when a *Diadasia* visits an unusual flower. (*D. enavata*, an oligolege on *Helianthus* which is also a host of *Nemognatha*, is an exception to this statement, but its nests have not been well studied.)

Three species of *Hornia* are associated with members of the genus *Anthophora* which nest in aggregations. E.G.L. and J.W.MacS. have observed aggregations of as few as 6 burrows where these beetles were present, and much larger ones where they were absent. Their initial presence in a starting aggregation must be largely fortuitous, but their adult habits would preclude any regular association with solitarily nesting species of *Anthophora*. The adults of *Hornia* are wingless and take no food. The female is fertilized in the bee cell in which she developed and deposits her eggs in the same cell. First-instar larvae are active at the time the first adult bees emerge. They attach themselves to these bees and thus infest a high percentage of the cells provisioned early in the season. Presumably because of this mechanism, this meloid is commonly present at a significant level in comparison to populations of its host.

Our knowledge of the habits of *Tetraonyx* is very incomplete. One species has been recorded as a parasite of *Centris*. Another was recorded as ovipositing on a plant which was being visited by a species of *Diadasia*. We obtained one adult from a cell of *Melitoma*. The host relationship may be similar to that of *Nemognatha*.

Hosts and partial life histories of two European species of *Zonitis* (*Z. immaculata* and *Z. praeusta*) are known; they ordinarily develop in nests of anthidiine, megachiline, and osmiine Megachilidae (Cros, 1928). Nothing was known of the larval habits of North American members of the genus until Selander and Bohart (1954) provided an account of *Z. atripennis flavida*, a parasite in the cells of the halictid bee *Nomia melanderi melanderi* in Utah. The habits of *Z. atripennis flavida* are similar to those of *Nemognatha*. Eggs are laid on stems, florets, and green pods of *Cleome*, and hatching first-stage larvae move to the vicinity of nectaries in open flowers, where they attach themselves to the visiting bees and are carried to the burrows. Each ultimately develops in a freshly provisioned cell, presumably after destroying the egg of the host. The *Zonitis* in nests of *Anthophora flexipes* (Table 5) was *Z. hesperis*; only 8 in a sample of 300 cells contained immature stages of this beetle.

Rhipiphorid beetles of the genus ***Rhipiphorus*** have been reared from cells of several genera of solitary bees. Limited evidence suggests that species of this genus are restricted to bee hosts belonging to particular genera. The adults are both weak fliers and short-lived. Females mate near the burrow from which they emerge and oviposit in the buds of specific host plants in the vicinity. The larvae attach themselves to visiting bees and are carried to cells being provisioned, where they enter the egg or larva of the bee. Later they emerge from the mature bee larva and consume it. Of the bee genera discussed here, only species of *Diadasia* have been recorded as hosts of ***Rhipiphorus***. Both *Diadasia* species that do not aggregate and those that do are involved. In the former case, the association appears to be due to the bee and the beetle utilizing a single plant group, and the relative abundance of ***Rhipiphorus*** is low. With *Diadasia* which nest in aggregations, the flower relationships are similar, and mating over the nest site would appear to favor the reproduction of the beetles. In these cases the beetles may destroy a much higher percentage of the host progeny.

Bombyliid flies are associated with all of the bee species shown in Table 5. They are absent from some aggregations, destroy less than 5% of the bee larvae in most populations, but have been recorded from more than 40% of the non-moldy cells in one sample (*Diadasia bituberculata*). Several possible explanations occur to us. Excepting the unusual probable oviposition habit of the bombyliid flies infesting cells of *M. marginella* in Chiapas, species of *Anthrax, Villa,* and *Heterostylum* oviposit while in flight by "throwing" eggs into the burrows of bees. Available data suggest that the position of the burrow entrance—i.e., in a vertical or horizontal substrate—and the presence or absence of an entrance tube or turret projecting beyond the soil surface, play important roles in the economy of these flies.

In one locality where both *A. linsleyi*, which makes no turrets, and the turret-making *A. bomboides stanfordiana* occupied the same vertical cliff and were attacked by the same species of fly, more than 4% of the cells of the first species, and only 2% of those of the second, contained this fly. This suggests that some degree of protection is furnished by the entrance tubes.

The two species of *Diadasia* listed in Table 5, like others of this genus, construct turrets above the burrow entrances in flat ground. These turrets, unlike those of many anthophorids, are delicate, and easily destroyed by water or other agencies. Within the genus there are two types of turrets (Eickwort, Eickwort, and Linsley, 1977): In some species they are erect chimneys with the entrances opening upward; in other species the turret is curved near its apex and the opening is lateral. MacSwain (unpublished) found parasitism to be low when curved turrets are present; it is apparently confined mainly to those burrows with turrets partially or completely destroyed. Species that make erect turrets appear to support proportionately higher populations of flies. The curved type is comparable to the cliff-dwelling species, in that oviposition involves directing eggs horizontally into an opening. While this evidence appears to suggest a function for *Diadasia* turrets in lowering the frequency of parasitization by bombyliids (see also Eickwort, Eickwort and Linsley, 1977), they may also have other functions, such as keeping loose dirt out of the burrows.

Present evidence on anthophorid bees does not allow evaluation of the effect of aggregation on population levels of bombyliids. Yet it may be noted that the majority of anthophorid species which construct turrets are those which nest in aggregations.

Bohart, Stephen, and Eppley (1960) note that *Nomia melanderi*, with most of its burrows in aggregations, all without turrets, may have more than 90% of its progeny destroyed by bombyliids in a heavily populated nest site, while scattered solitary burrows in the same area escape such heavy attack.

Mutillid wasps have been associated with species of the three genera of anthophorid bees considered here. Brothers (1972) summarizes much of what is known of the biology of this family. Adults of *Sphaeropthalma* are chiefly nocturnal, and those of *Dasymutilla* diurnal. Many species have hymenopterous hosts, but none is known to be host-specific. The females of these wasps search their environment for the burrows of bees and wasps. Upon entering an anthophorid burrow, the mutillid digs along the sides of cells containing mature host larvae and oviposits through a pit excavated in the side of each host cell. The hard substrates utilized by the anthophorid species considered here may reduce the efficiency of mutillids and explain their relative low frequency in these populations. A significant discovery on the habits of a species of *Sphaeropthalma* has been contributed by Ferguson (1962). He found that this mutillid completes repeated generations on the progeny of the host bee (*Diadasia*) produced in one generation. Since the larvae of this bee may remain in diapause during several unfavorable years, the cumulative effect of the wasp may be considerable. Further, the large aggregations of burrows of this bee may facilitate the life of the parasite, because many potential hosts are available in a limited area.

A mantispid species of the genus *Plega* was associated with cells of *Melitoma* in all three of the large aggregations (sites 12–14) in the state of Chiapas, Mexico. From this it might be inferred that the mantispid is host-specific and that its habits, if better known, would show adaptations for the maintenance of such an association. In the absence of such data, it should be noted that there are few records of mantispids from bee cells. Other host records of *Plega* species in North America are from moth and beetle pupae, but also from cells of a megachilid bee (Parker and Stange, 1965). These authors recognize two groups of *Plega*: one parasitic on bees and wasps, the other on subterranean insects.

Several species of torymid wasps of the genus *Monodontomerus* have been found in association with cells of both *Anthophora* and *Melitoma*. Rau (1926) reported a species of *Monodontomerus* as extremely abundant about a bank that contained a mixed population of *Anthophora* and *Melitoma*. With this exception, observations of *Monodontomerus* have shown it to be frequent, but accounting for the destruction of a low percentage of bee larvae. Compensating for this, in part, is the fact that 35 or more wasps may emerge from a single host cell. The factor limiting their abundance may be failure of the female to dig alongside the serially arranged cells and oviposit through the thick cell walls. When a mutillid wasp tunnels beside a group of cells and, moreover, excavates a pit in the side of each cell, entry by *Monodontomerus* is facilitated. In one study it was shown that 0.5% of the bee larvae had been destroyed by the chalcidoid and 2.5% by the mutillid. Of the latter, however, 67% had been hyperparasitized by the chalcidoid.

Parasitic bees have been reared from the cells of a number of species of *Anthophora* and of *Diadasia* (not from the species listed in Table 5), but not from those of *Melitoma*. Data on the relative abundance of host and parasitic bees suggest that the parasitic species are usually a minor drain on their host's population. Among 18 tabulations of numbers and species of insects from *Anthophora* cells, parasitic bees were encountered in 17 samples. In 14 of these they occupied from 1.4% to 7.4% of the cells, but in the remaining 3 samples, *Melecta* parasitized 21% to 33% of the cells.

Ashmead (1894; and see Robertson, 1899) apparently saw the nomadine bee *Triepeolus donatus* around or entering *Melitoma* nests at site 25 (Utica, Mississippi). This *Triepeolus* ranges farther north than *Melitoma*; if it parasitizes *Melitoma* at all, it must also have other hosts. Bertoni (1929) recorded *Coelioxys paraguayensis* (?) entering *Melitoma* nests at site 5 (Puerto Alegre, Paraguay). While most *Coelioxys* species attack megachilids, there are Old World species that attack *Clisodon* and *Anthophora*, and parasitization of *Melitoma* is at least possible.

Scavenger insects such as *Trogoderma* appear in bee aggregations that have persisted over many generations. Since long survival is characteristic of aggregations of several species of *Anthophora*, the greatest abundance and largest number of kinds of scavengers have been recorded from *Anthophora* nest sites. *Melitoma* aggregations appear to build up rapidly, in a few generations, and then disappear. As a result, scavenger species are uncommon.

BEES ASSOCIATED WITH *MELITOMA* NESTS

In addition to the organisms discussed above that are associated with *Melitoma* cells and usually destructive to their contents, there are several species of bees that nest in abandoned burrows of *Melitoma* or among *Melitoma* nests. None of these species is known to destroy *Melitoma* cells; their effect on the *Melitoma* population is probably nil.

Chalicodoma abacula

Several examples of the megachilid *Chalicodoma* (*Chelostomoides*) *abacula* emerged from cells of *Melitoma* at site 14 (Francia, Chiapas). Only one individual was reared from each cell utilized. In this area the species was parasitized by *Coelioxys chichimeca* and an undescribed species of a genus near *Eusapyga*, both of which were reared from cells which had contained *Chalicodoma abacula.* At site 12 (Tuxtla Gutiérrez) Selander found 2 of 13 cells appropriated by this species to contain two *Chalicodoma* cells each. Two *Chalicodoma* cells also contained dead fifth-instar larvae of an unidentified lyttine meloid. Adults in Selander's sample began emerging near the middle of November in the laboratory.

Dianthidium (Dianthidium) curvatum

At site 25 (Lebanon, Missouri), this megachilid bee nested in abandoned cells and burrows of *Melitoma taurea* (Michener, 1975). It was originally misidentified as *Paranthidium jugatorium*, but its identity is shown by voucher specimens in the Snow Etomological Museum, University of Kansas. At least 9 of about 40 old nests were occupied by *Dianthidium*. Since most megachilids utilize a variety of pre-existing burrows, it seems unlikely that *D. jugatorium* is restricted to old *Melitoma* nests.

Dianthidium (Epanthidium) bertonii

This megachilid appropriated several *Melitoma* burrows for its nests at site 5 in Paraguay (Bertoni, 1929).

Centris

A number of specimens of *Centris trigonoides* (= *hoplopoda*; see Moure, 1960) were reared from cells of *Melitoma marginella* from sites 12–15 (Chiapas). This species apparently appropriates used cells rather than excavating its own. Whether this is true regularly or throughout the range of the species remains to be determined. At site 14 (Francia) the *Centris* was parasitized by the bombyliid fly *Anthrax mexicana*, and by the bee *Mesocheira bicolor.* Examples of both parasites were reared, and others were captured while flying about the nesting area early in April.

A single individual of *Centris totonaca* was also reared from an appropriated *Melitoma* cell at site 14, and numerous cocoons of this *Centris* were found in old *Melitoma* cells at site 20 (Culiacán, Sinaloa, Mexico—Parker, 1977). The parasitic bee *Mesocheira bicolor* was reared from a cell of *C. totonaca*. A third species of *Centris* was present at site 13 (Rio La Venta, Chiapas).

Bertoni (1929) has recorded *C. lanipes* nesting with *M. segmentaria* in Paraguay. In some localities the *Centris* was more abundant than the *Melitoma*, while at others the relationship was reversed. Apparently both species make burrows; Bertoni says nothing about the *Centris* appropriating cells or burrows of *Melitoma.*

Anthophora

At site 27 (Wicks, Missouri), Rau (1926, 1929, 1934) studied a sheltered bank and adjacent flat ground inhabited primarily by *Anthophora abrupta*, but with some nests of *A. raui* and of *Melitoma*. The *Melitoma* were largely in separate areas, but there was some intermixture of nests.

Ancyloscelis apiformis

This species, formerly called *A. armata*, is a small anthophorid which, like *Melitoma,* collects pollen from *Ipomoea*. Also, like *M. segmentaria* and *M. marginella*, it ranges from southernmost Texas to Argentina, with very close relatives extending north in the plains (as far as Colorado). Its nests have been studied at 4 sites (3, 9, 10, and 19) from Uruguay to Mexico. At each site its burrows and cells were intertwined and intermixed with those of *Melitoma*, bees of each genus sometimes encroaching upon old cells of the other. Indeed, Brèthes (1909) suggested that one might be a parasite of the other (which is not the case), and Michener (1974a) suggested that preparation of a site by an aggregation of *Melitoma* might be necessary for successful nesting by *Ancycloscelis*.

References

Ashmead, W. H.

1894. The habits of the aculeate Hymenoptera—I. Psyche, 7: 19-26.

Beal, R. S., Jr.

1954. Biology and taxonomy of the Nearctic species of *Trogoderma* (Coleoptera: Dermestidae). Univ. California Publ. Entomol., 10: 35-102.

Bertoni, A. de W.

1918. Notas entomológicas (biologicas y systemáticas). An. Cient. Paraguayos, 2: 219-232.

1929. Nidification en colonias de abejas antofóridas. Rev. Soc. Cient. Paraguay, 2: 223-224.

Bodkin, G. E.

1918. Notes on some British Guiana Hymenoptera (exclusive of the Formicidae). Trans. Entomol. Soc. London, 1917: 297-321.

Bohart, G. E., W. P. Stephen, and R. K. Eppley.

1960. The biology of *Heterostylum robustum* (Diptera: Bombyliidae), a parasite of the alkali bee. Ann. Entomol. Soc. Amer., 53: 425-435.

Brèthes, J.

1909. Una anthophorina Parásita? An. Mus. Nac. Buenos Aires, (3)12: 81-83.

Brothers, D. J.

1972. Biology and immature stages of *Pseudomethoca f. frigida*, with notes on other species. Univ. Kansas Sci. Bull., 50: 1-38.

Casey, T. L.

1915. A review of the American species of Rutelinae, Dynastinae and Cetoninae. Memoirs on the Coleoptera, 6: 1-394.

Chandler, L.

1958. The biology of *Melitoma taurea* (Say) in Indiana (Hymenoptera, Anthophoridae). Bull. Entomol. Soc. Amer., 4: 101 (abstract).

Cockerell, T. D. A.

1913. Descriptions and records of bees—XLIX. Ann. Mag. Nat. Hist., (8)11: 185-195.

1934a. Records of western bees. Amer. Mus. Novitates, 697: 1-15.

1934b. The wild bees. Natural History, 34: 748-753.

Cockerell, T. D. A., and W. Porter.

1899. Contributions from the New Mexico Biological Station—VII. Observations on bees, with descriptions of new genera and species. Ann. Mag. Nat. Hist., (7)4: 403-421.

Cole, F. C.

1957. New bombyliid flies from Chiapas, Mexico (Diptera). Pan-Pac. Entomol., 33: 200-202.

Corporaal, J. B.

1950. Cleridae. *In* Coleopterorum Catalogus Supplementa, 23 (ed. 2): 5-73.

Cros, A.
1928. Études biologiques sur les *Zonitis* (Meloidae). *In* Encyclopédie Entomologique, Série B. Coleoptera, 3: 7-34, pl. 1.

Doello-Jurado, M.
1912. Apuntes entomologicos. Nidification y habitos de una abeja sylvestre, la *Entechnia.* Bol. Soc. Physis, 1: 52-56.

Eberhard, W. G.
1974. The natural history and behaviour of the wasp *Trigonopsis cameronii* Kohl. Trans. Roy. Entomol. Soc. London, 125: 295-328.

Eickwort, G. C., K. R. Eickwort, and E. G. Linsley.
1977. Observations on nest aggregations of the bees *Diadasia olivacea* and *D. diminuta.* Jour. Kansas Entomol. Soc., 50: 1-17.

Fattig, P. W.
1943. The Mutillidae or velvet ants of Georgia. Emory Univ. Mus. Bull., no. 1: 1-24.

Ferguson, W. E.
1962. Biological characteristics of the mutillid subgenus *Photopsis* Blake and their systematic values (Hymenoptera). Univ. California Publ. Entomol., 27: 1-92.

Gahan, A. B.
1941. A revision of the chalcid-flies of the genus *Monodontomerus* in the United States National Museum. Proc. U. S. Nat. Mus., 90: 461-482.

Gorham, H. S.
1882. Coleoptera. *In* Biologia Centrali-Americana, 3, part 2, xii + 372 pp., 13 pl.

Hobbs, G. A., W. O. Nummi, and J. F. Virostek.
1961. *Anthophora occidentalis* Cress. (Hymenoptera: Apidae) and its associates at a nesting site in southern Alberta. Canadian Entomol., 93: 142-148.

Horn, G. H.
1880. Contributions to the Coleopterology of the United States, no. 3. Trans. Amer. Entomol. Soc., 8: 139-154.

Hungerford, H. B., and F. X. Williams.
1912. Biological notes on some Kansas Hymenoptera. Entomol. News, 23: 241-260.

Hurd, P. D., Jr.
1951. The California velvet ants of the genus *Dasymutilla* Ashmead (Hymenoptera: Mutillidae). Bull. California Insect Survey, 1: 89-118.

Krombein, K. V.
1951. Superfamily Scolioidea. *In* Muesebeck, Krombien, and Townes, Hymenoptera of America north of Mexico. U. S. Dept. Agric. Monograph, 2:735-778.

Linsley, E. G.
1960. Observations on some matinal bees at flowers of *Cucurbita, Ipomoea,* and *Datura* in desert areas of New Mexico and southeastern Arizona. Jour. N. Y. Entomol. Soc., 68: 13-20.

Linsley, E. G., and J. W. MacSwain.
1942a. Bionomics of the meloid genus *Hornia* (Coleoptera). Univ. California Publ. Entomol. 7: 189-206.
1942b. The parasites, predators and inquiline associates of *Anthophora linsleyi.* Amer. Midland Nat., 27: 402-417.
1952a. Notes on the biology and host relationships of some species of *Nemognatha* (Coleoptera: Meloidae). Wasmann Jour. Biol., 10: 91-102.
1952b. Notes on some effects of parasitism upon a small population of *Diadasia bituberculata* (Cresson) (Hymenoptera: Anthophoridae). Pan-Pac. Entomol., 28: 131-135.
1955. Two new species of *Plega* from Mexico (Neuroptera, Mantispidae). Pan-Pac. Entomol., 31: 15-19.

1957. The nesting habits, flower relationships, and parasites of some North American species of *Diadasia* (Hymenoptera: Anthophoridae). Wasmann Jour. Biol., 15: 199-235.

Linsley, E. G., J. W. MacSwain, and R. F. Smith.

1952. The bionomics of *Diadasia consociata* Timberlake and some biological relationships of emphorine and anthophorine bees (Hymenoptera, Anthophoridae). Univ. California Publ. Entomol., 9: 267-290.

1956. Biological observations on *Ptilothrix sumichrasti* (Cresson) and some related groups of emphorine bees. Bull. So. California Acad. Sci., 55: 83-101.

MacSwain, J. W.

1958. Taxonomic and biological observations on the genus *Hornia* (Coleoptera: Meloidae). Ann. Entomol. Soc. Amer., 51: 390-396.

Marston, N.

1970. Taxonomic study of the known pupae of the genus *Anthrax* (Diptera: Bombyliidae) in North and South America. Smithsonian Contr. Zool., no. 100: 1-18.

Michener, C. D.

1974a. Further notes on nests of *Ancyloscelis* (Hymenoptera: Anthophoridae). Jour. Kansas Entomol. Soc., 47: 19-22.

1974b. The social behavior of the bees. Cambridge, Mass.: Harvard Univ. Press, 404 pp.

1975. Nests of *Paranthidium jugatorium* in association with *Melitoma taurea* (Hymenoptera: Megachilidae and Anthophoridae). Jour. Kansas Entomol. Soc., 48: 194-200.

Michener, C. D., and R. B. Lange.

1958. Observations on the ethology of Neotropical anthophorine bees (Hymenoptera, Apoidea). Univ. Kansas Sci. Bull., 39: 69-96.

Michener, C. D., R. B. Lange, J. J. Bigarella, and R. Salamuni.

1958. Factors influencing the distribution of bees' nests in earth banks. Ecology, 39: 207-217. (Also published in Portuguese, in Dusenia, 8: 1-24.)

Mickel, C. E.

1928a. Biological and taxonomic investigations on the mutillid wasps. Bull. U. S. Nat. Mus., no. 143: 1-351.

1928b. The biotic factors in the environmental resistance of *Anthophora occidentalis* Cress. (Hym.: Apidae; Dip., Coleop.). Entomol. News, 39: 69-78.

Moure, J. S.

1960. Notes on the types of Neotropical bees described by Fabricius. Studia Ent. [Rio de Janeiro], 3: 97-160.

Parker, F. D.

1977. Biological notes on some Mexican bees. Pan-Pac. Entomol., 53: 189-192.

Parker, F. D., and L. A. Stange.

1965. Systematic and biological notes on the tribe Platymantispini and description of a new species of *Plega* from Mexico. Canadian Entomol., 97: 604-612.

Porter, J. C.

1951. Notes on the digger-bee, *Anthophora occidentalis*, and its inquilines. Iowa State Coll. Jour. Sci., 26: 23-30.

Rau, P.

1926. The ecology of a sheltered clay bank: a study in insect sociology. Trans. Acad. Sci. St. Louis, 25: 159-276.

1929. The biology and behavior of mining bees, *Anthophora abrupta* and *Entechnia taurea*. Psyche, 36: 154-181.

1934. Notes on the behavior of certain solitary and social bees. Trans. Acad. Sci. St. Louis, 28: 219-224.

1947. Bionomics of *Monodontomerus mandibularis* Gahan, with notes on other chalcids of the same genus. Ann. Entomol. Soc. Amer., 40: 221-226.

Robertson, C.

1899. Flower visits of oligotropic bees. Bot. Gaz., 28: 215.

1914. Origin of oligotropy of bees (Hym.). Entomol. News, 25: 67-73.

Sallé, A.

1833. Observations sur les moeurs de plusieurs Coléoptères du Mexique. Revue Entomol. (Silberman), 1: 237-243.

Say, T.

1837. Descriptions of new North American Hymenoptera, and observations on some already described. Boston Jour. Nat. Hist., 1: 210-305, 361-416.

Schrottky, C.

1902. Hymenoptères nouveaux de l'Amérique méridionale. An. Mus. Nac. Buenos Aires, 7: 309-314.

Selander, R. B.

1960. Bionomics, systematics, and phylogeny of *Lytta*, a genus of blister beetles (Coleoptera, Meloidae). Illinois Biol. Monographs, 28: 1-295.

Selander, R. B., and G. E. Bohart.

1954. The biology of *Zonitis atripennis flavida* LeConte (Coleoptera, Meloidae). Wasmann Jour. Biol., 12: 227-243.

Thorp, R. W.

1969. Ecology and behavior of *Anthophora edwardsii*. Amer. Midland Nat., 82: 321-337.

Torchio, P. F.

1971. The biology of *Anthophora (Micranthophora) peritomae* Cockerell. Los Angeles County Mus. Contr. Sci., 206: 1-14.

1974. Notes on the biology of *Ancyloscelis armata* Smith and comparisons with other anthophorine bees. Jour. Kansas Entomol. Soc., 47: 54-63.

Torchio, P. F., and N. N. Youssef.

1968. The biology of *Anthopohra (Micranthophora) flexipes* and the cleptoparasite, *Zacosmia maculata*, including a description of the immature stages of the parasite. Jour. Kansas Entomol. Soc., 41: 289-302.

FIGURES

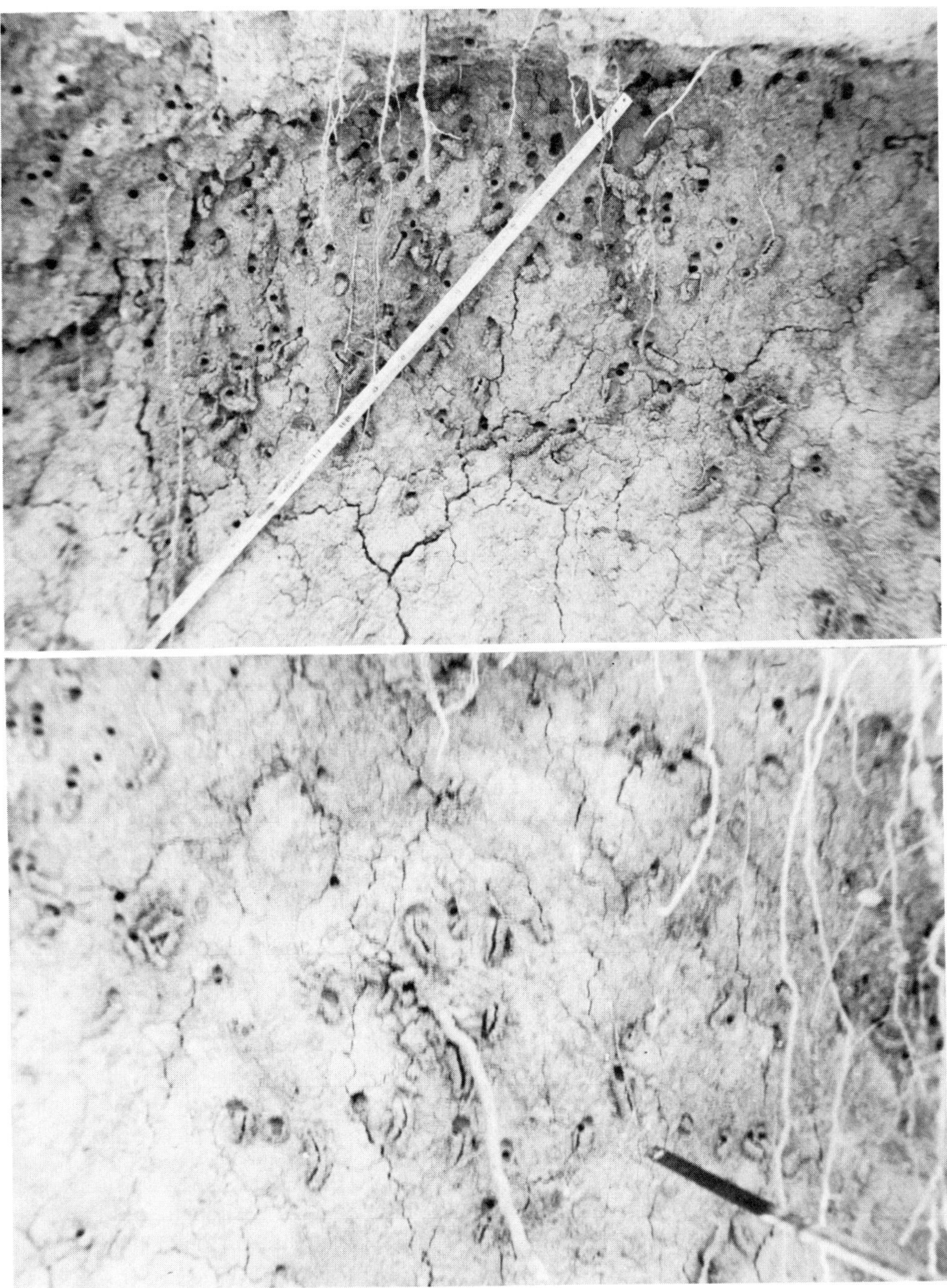

FIGURE 1. **Above.** Portion of a bank near Soyalo (site 15) showing nests of *Melitoma marginella* concentrated beneath a weak overhang from which rootlets are hanging. The visible part of the ruler is 45 cm long. **Below:** Portion of the bank showing some well-formed entrance tubes of *M. marginella*, each divided on the outer surface by a fissure.

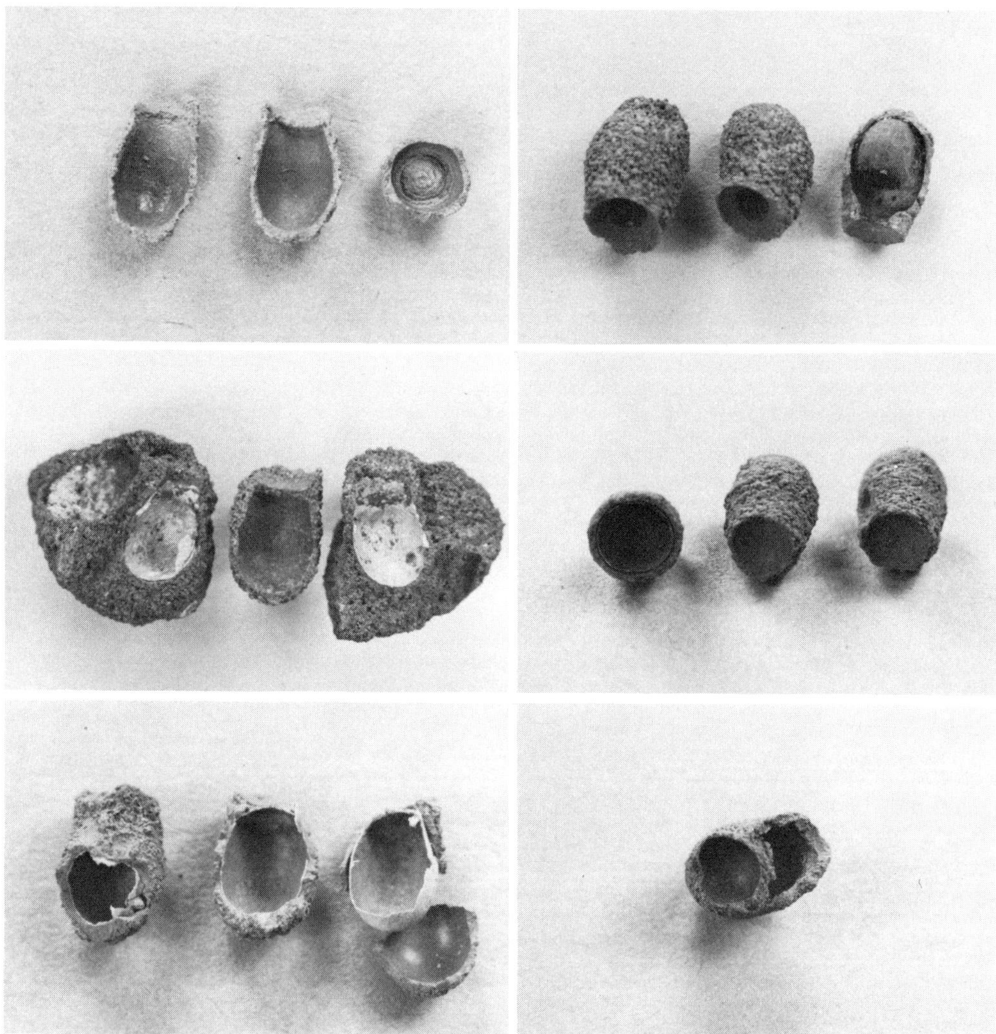

FIGURE 2. Cells of *Melitoma.* **Top left:** Two cells of *M. segmentaria* from site 9 (Cali, Colombia), split longitudinally. Both had been provisioned and closed; the provisions were removed to show the shining inner surface secreted by the adult female bee. The spiral pattern of the inner surface of the cell closure is shown at the right. **Top right:** Cells of *M. marginella* from site 14 (Chiapas, Mexico). The first two show variation in cell size and emergence holes of parasites, the third is split to show the opened cocoon of *Plega melitomae.* **Center left:** Cells of *M. segmentaria* from site 10 (Kourou, French Guiana). Note that some did not separate from the matrix but that the middle one did so, as is usual for *Melitoma* cells. The white linings are thin layers of larval fecal material plus the thin cocoon spun by the mature larva. **Center right:** Cells of *M. marginella* from site 14 (Chiapas, Mexico). The cross-section shows the thin fecal layer and cocoon slightly separated from the mud cell wall. The two intact cells show the smooth, concave outer surfaces of the cell closures. **Bottom left:** Cells of *M. taurea* from site 26 near Lebanon, Missouri, broken to show the rigidity of the thin white layer of larval feces plus the cocoon. **Bottom right:** Cell of *M. segmentaria* from site 9 (Cali, Colombia), constructed partly within a pre-existing cell. The wall in the middle was built up in the cavity of the old cell, without substrate against which to shape the new wall.

FIGURE 3. **Left:** *Nemognatha chrysomeloides*, showing long mouthparts. **Right:** *Melitoma marginella*, female, showing long mouthparts.

FIGURE 4. *Cymatodera hopei.*

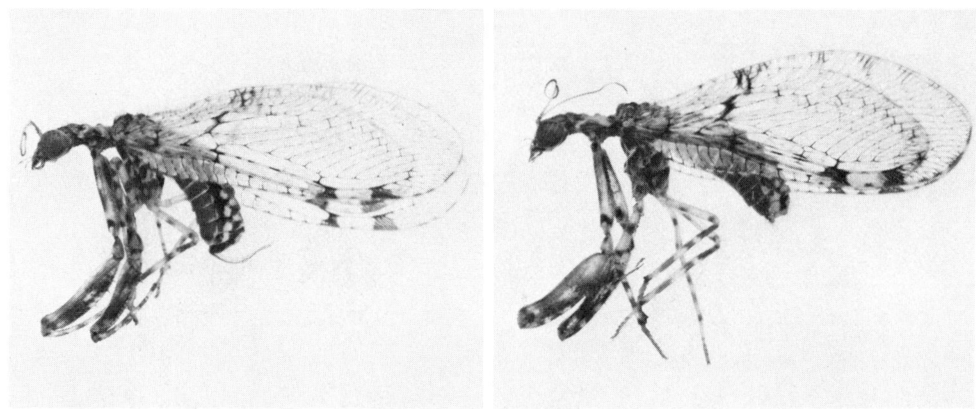

FIGURE 5. *Plega melitomae.* **Left,** female; **right,** male.

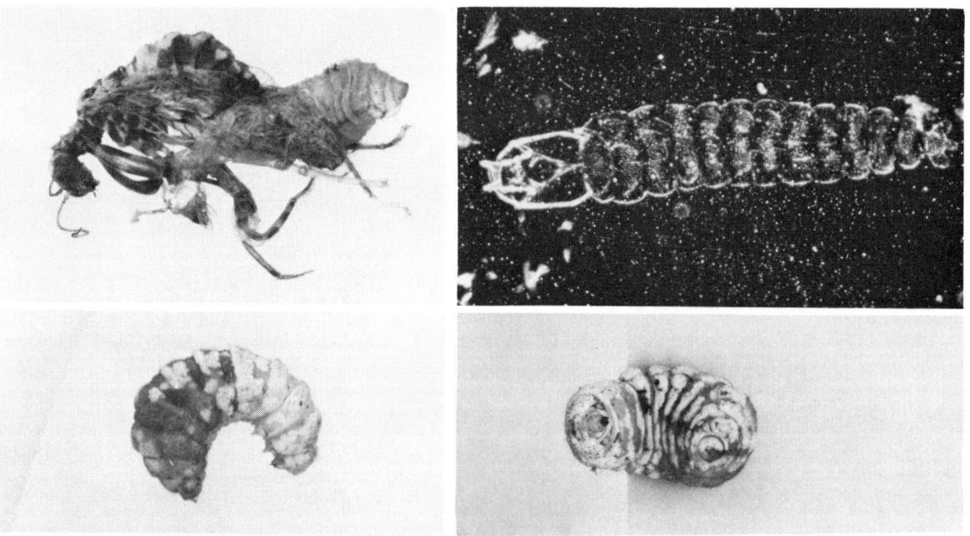

FIGURE 6. *Plega melitomae.* **Above:** adult emerging from pupal skin, and first stage larva. **Below:** two views of mature larva.

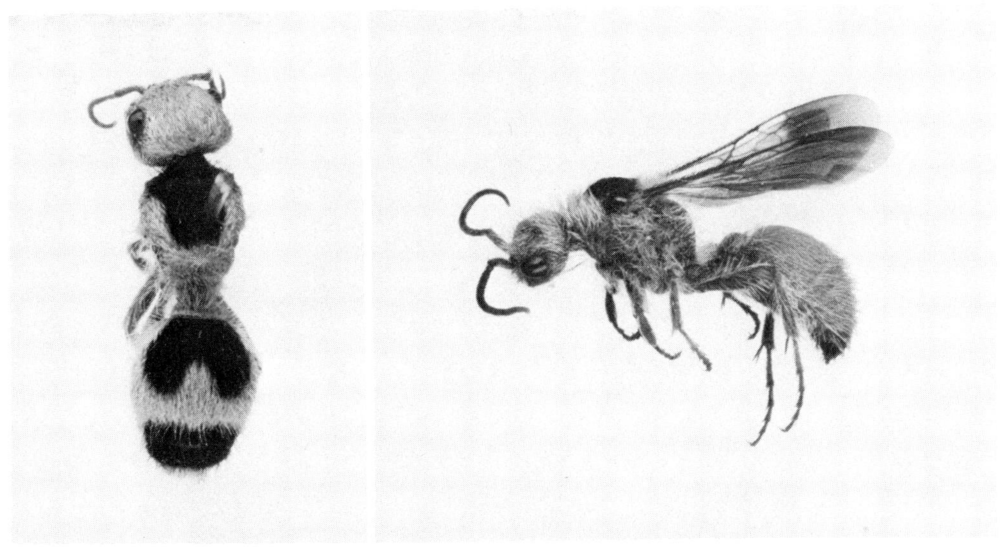

FIGURE 7. *Dasymutilla canina.* **Left,** female; **right,** male.

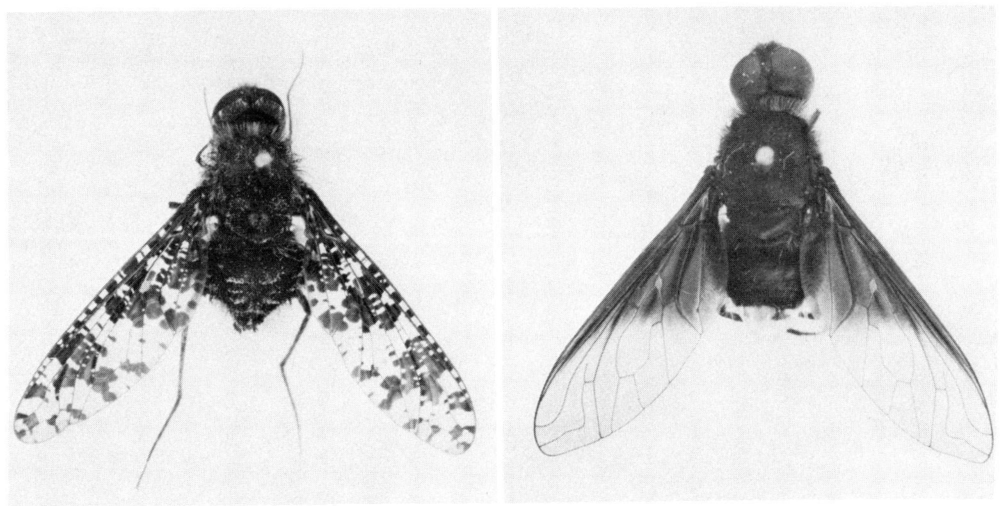

FIGURE 8. **Left:** *Anthrax cintalapa*; **right:** *A. mexicanus.*